Charcot's

Conversion Disorder & Functional Symptoms in Neurology

❧

By Simon Overton

About the Author

Simon Overton studied English and Philosophy at Bristol and York. He taught at Emmanuel College in Gateshead before moving to the Lake District to teach English and go climbing. Following a viral infection he was diagnosed as suffering from a Functional Neurological Deficit. Despite recovery he suffered major relapses climbing in the Tien Shan and the Swiss Alps. When the Head of Mental Health Services suggested further testing for organic illness this was ignored. Following the assistance of an internationally renowned expert in missed diagnoses, his doctors conducted further tests showing damage to his autonomic nervous system and orthostatically mediated weakness. Further neurological examination suggested by family history was indicative of hereditary spastic paraplegia. His family has a history of "stiff legs" and cardio/cerebro-vascular disease. His sister was also treated for orthostatically related central nervous system symptoms. He continues to teach in the Lake District and lives with his partner Fiona and their Dalmatian, Amy.

www.simonjoverton.co.uk

Edited by Christine Furmston

©S. Overton 2009

ISBN 978-1-4092-6542-9

For Byron Hyde

I would have done exactly the same as you and also be gnashing my teeth for being such a fool and at the same time contemplating when I would next try something ridiculously wonderful.

Byron Hyde MD

Charcot's Bad Idea

Conversion Disorder & Functional Symptoms in Neurology

Contents:

Preface
1. Portrait of a Diagnosis
2. Prognosis, Diagnosis and Treatment
3. A Brave New World
4. Alas, the Storm is Come Again
5. Studies of Hysteria
6. Mechanisms of Compliance
7. Hysteria as Failed Communication
8. An Unholy Alliance
9. The Case of Ean Proctor
10. Wessely, Sharpe and the Age of Steam
11. Conclusion
12. Acknowledgments
13. Index
14. Notes & Glossary

Preface

To understand where we are we must know where we have come from. Science and medicine are not bereft of those political, economic and historical forces that shape their thinking and praxis. Substantial medical evidence spanning many decades continues to be ignored by those in positions of power who are abusing that power to further their own vested interests. Once an idea takes root in the scientific community it is not reason but the ability to gather kudos and grab research funding that informs dominant thinking. These groups and individuals are helped immeasurably by the creation of bogus disease categories, categories such as 'Chronic Fatigue Syndrome' and other cultural rather than scientific concepts such as somatisation disorder, functional weakness, conversion disorder, hysteria, and so on.

Despite an utter lack of scientific legitimacy and the enormous costs to the many patients involved and to the community at large, these massive medical frauds have continued in perpetuity almost entirely unchallenged by the world's media, human rights groups, and governments. It is a worldwide disgrace and one of the biggest scandals in the history of medicine.

How much more extreme do the suffering and abuse caused by these malign scams have to be? How many more very ill patients have to be denied even basic medical care? How many more hundreds of thousands of children and adults worldwide have to be left severely disabled or dead through inappropriate treatment?

Simon Overton is to be congratulated for creating such a timely, intelligent and compelling book on this important topic. More uncompromising educational efforts like this one must be produced if change is ever to occur. The fox has been left in charge of the henhouse for far too long already. Knowledge is power.

May the day soon come when such books (and other advocacy projects) are no longer needed, and when patients can rely on something as simple as treatment based on legitimate scientific evidence and on the reality of their pathology – rather than being subjected to various self-serving and illogical pseudo-scientific 'theories'. Such ideas are extremely unlikely to help any of the patient groups involved to regain their health. It is not illness beliefs within the body of the patient that must be treated but rather ideologies within some sectors of the medical profession.

Jodi Bassett: February 2009, Perth, Australia.
(Author of "A Hummingbird's Guide to ME")

Variability is the law of life, and as no two faces are the same, so no two bodies are alike, and no two individuals react alike and behave alike under the abnormal conditions which we know as disease. Medicine is a science of uncertainty and an art of probability.
William Osler, Canadian Physician

1) Portrait of a diagnosis

"Science must begin with myths and with the criticism of myths"- Karl Popper

Patients often present to doctors with illness for which there is no obvious organic explanation despite investigation. Historically medicine lacks a tradition of ignorance and has a tendency to suggest explanations for illness and treat the patient accordingly, even when these explanations and cures are bizarre and absurd. Examples in the medical canon range from the leeches and bloodletting of the Ancient Greeks, to the lobotomy of pre 1970's neurosurgery. Illness that is not easily explained can challenge the hegemony of medicine. Patients and others can see the psychiatric referral that often follows as a means of controlling this threat to the physician's authority, for how after millennia of "progress" can medicine itself be deviant or deficient? Perhaps as a result of this tradition some contemporary thinkers in neuropsychiatry, (Manu, Sharpe, Wessely and others) place the blame on the patient for their illness, or more specifically the ideas held by a patient about their illness[1]. In relation to "Chronic Fatigue Syndrome," Sharpe writes:

Inaccurate and unhelpful beliefs, ineffective coping behaviour, negative mood states, social problems, and pathophysiological processes all interact to perpetuate the illness.[2]

It is in this curious statement that a key argument of this book can be found. For in effect what Sharpe is doing is privileging unexplained somatic symptoms as distinct from classical disease entities. A view that can also be found in the National Institute for Clinical Excellence (NICE) championing psychiatric therapies, such as cognitive behavioural therapy, for chronic fatigue syndrome but not for multiple sclerosis. It is natural to assume that unhelpful ideas might be tackled in coping with illness. It is obvious also that social circumstances do have a profound impact on the prognosis of disease, especially in emerging economies.

But what Sharpe is claiming is in reality something quite special. He claims that the ideas held by a patient can actually perpetuate the course of an illness. It is not through treating or diagnosing the pathophysiological processes that a patient might get better but by denouncing their own belief system and coping mechanisms in favour of those ideologies Mike Sharpe, the therapist or other physician deem to be superior. If the patient does not improve then the question must be asked, if organicity cannot be blamed then who or what is to blame? That this system of ignoring pathology could actively denounce organic investigation in favour of a talking cure was given voice by one of Sharpe's colleagues, Prof. Peter White, in a "You and Yours" broadcast on "Chronic Fatigue Syndrome" transmitted Monday 5th November 2007:[3]

> *Interviewer: "You mentioned tests that you don't think it's right for you to do, such as...?"*
> *Prof. Peter White: "Such as the tilt table test – I would have to exclude a condition called POTS (where the blood pressure falls on standing up [sic]). I don't think that's justified."*

> *Interviewer: "So you think they're unethical because they're too demanding?"*
> *Prof. Peter White: "Yes."*

In a world where funding for public health care is under increasing pressure, such ideologies have enormous political appeal.

Medicine cannot escape from the reality that within a publically funded health care system there is an economic and ideological pressure on doctors to suggest that any testing is extensive and patients are appropriately investigated. There is also a lack of recognition that all illness, from the presentation of the patient to the actual interpretation of alleged "illness behaviour" is influenced by social and cultural factors. The assumption that medically unexplained symptoms are less organic and more psychosocial than other illnesses is one concept that this book will investigate.

Faced with complex neurological symptoms for which no obvious disease process could be found, physicians such as Jon Stone and Mike Sharpe have pioneered methods of diagnosing and treating this group of patients, in a bid to improve their management and care. This is against a social backdrop where the ascendency of routine neurological investigations, such as MRI, has eclipsed trust in a physician to diagnose illness purely on professional judgement. Widespread belief in the omnipotence of MRI and what is actually very low-resolution technology (normally 1.5 tesla compared to a currently possible 9 tesla, e.g. 1.5 megapixels compared to a 9 megapixel camera) is profound. Yet gross disability can exist with normal MRI, hereditary spastic paraplegia and its variants, for example.

The work of Stone and Sharpe is also against a cultural backdrop where medicine is increasingly expected to give biochemical explanations for everything from anxiety to senility. Focusing on a patient's experience of symptoms rather than pathology, Stone, Sharpe and others follow in the footsteps of Jean-Martin Charcot, a 19th century French neurologist and colleague of Sigmund Freud. In a development of Charcot's later theories they suggest that it is the patient's own ideas about themselves or their bodies that have made them ill. That everyday sensations can make: "them worried about a stroke, multiple sclerosis etc."[4] and thus blossom into theatrical interpretations of these disease entities. They suggest the patient's initial idea has somehow fundamentally altered their biochemistry and brain function, so that as Stone claims:

> *The mechanism of [the patient's] weakness could be seen as residual hemi-depersonalisation or a disconnection between the brain/mind and one half of the body.*[5]

Whether it is possible for half of me to be disconnected from the other half prompted by an idea is, some would argue, a controversial point. The body is not composed of discrete units but rather interdependent organisms. What Stone is therefore postulating is the notion that I can disconnect myself from myself. That intellectual activity can cause not just momentary lapses as I forget my sunglasses are on my head, but profound deficits in which no matter how hard I look for them, I am effectively never able to find them. Extended to hysterical paralysis the logic dictates that no matter how exhaustive or indifferent I am in the search for my legs, I am unable to find the mechanisms to move them, and this proceeds not from biochemical anomalies, but from mysterious psychological mechanisms; "hemi-depersonalisation".

In making such a diagnosis acceptable to the patient, old linguistic frameworks have been adopted and re-worked in the laudable effort to improve the care of a group of patients, a group often considered a waste of time by many neurologists.

Charcot was born in Paris in 1825, the eldest of four sons of a local carriage builder. In 1882 he established a neurology clinic at Salpêtrière, the first of its kind in Europe. It was here, at a time when medicine was blissfully bathed in ignorance, Charcot began working on the problem of hysteria and medically unexplained symptoms. His work continues to reverberate throughout the increasingly frustrated and distressing experience many patients have in dealing with neurological services. As the Hon. Ian Gibson has stated, "This is a problem experienced by many people in the UK."[6]

In neurology, even today, patients who have unexplained symptoms account for around a third of a neurologist's workload[7]. This is the highest amongst medical disciplines. At the other end of the scale dermatology encounters unexplained symptoms in only 5%[8] of patients. Looking beneath the skin is the diagnostic feat that most disciplines of medicine must perform in order to work out what exactly is wrong with the patient. In the past many neurological patients, with negative tests and peculiar symptoms, would have been labelled as suffering from hysteria.

With the growth of patient choice and the tendency of any patient told they were hysterical to walk out of the consulting room, new terms have come into play. In this book we will focus on the growing use of the term "functional", developed because of a "need to develop constructive ways of talking with patients."[9]

This book is therefore not about the history of hysteria but as to how historical concepts of hysteria have influenced contemporary diagnosis and treatment. It is about how the work of Charcot came to be revisited in an effort to avoid acknowledging the complex biochemical basis of symptoms that against a backdrop of cultural expectation continue to embarrass modern medicine. It is about how many patients see themselves as presented with a sleight of hand rather than a diagnosis.

Despite claims of neutrality for the term "functional," many patients are still diagnosed using "positive signs" that traditionally are linked to hysteria. These are tests that the doctor might carry out in physically examining the patient. They involve checking for patterns of weakness in the arms and legs, reflexes and other more specific tests such as Hoover's sign. Thus from the outset, the focus is on physical signs that Freud fancifully and without recourse to systematic study emphatically linked to emotional rather than organic disturbance. That a sign, such as Hoover's, may proceed from simple misunderstanding or from co-morbid organic injury is an area of study as yet unexplored.

Hoover's sign was developed by Dr. Charles Franklin Hoover (1865-1927), a doctor in Cleveland Ohio. The technique was first described in the Journal of the American Medical Association in 1908. The article was entitled, "A new sign for the detection of malingering and functional paresis of the lower extremities." and was based on the astonishingly large evidence base of only four patients. That malingering and hysteria could be dealt with, in the same paper and by the same test, was a new development for a condition that Jean-Martin Charcot had taken pains to categorise as distinct from feigning.

In using Hoover's sign the person is asked to lift their weak leg. In organic hemiplegia this allegedly causes all patients to involuntarily exert counterpressure with the heel of the opposite leg. The assumption is that patients will not brace their good leg to support the weakened limb. In a patient with hemiplegia caused by a lesion within the brain the examiner feels pressure on the hand beneath the healthy leg. In hysterical hemiplegia, or weakness not caused by a lesion, no pressure is felt. Hoover's work in refining tests for hysteria stemmed from his dissatisfaction with the work of Joseph Babinski and his very similar enterprise of defining psychopathology through physical signs.

Babinski was born in 1857 and had worked under the direction of Charcot at the Salpetriere hospital in Paris. Charcot had developed a powerful hold on all thinking about hysteria, a hold that for the sake of their careers few French neurologists and psychiatrists dared to challenge. Charcot held that hysteria was hereditary and that particular signs could be found by examining the neurotic patient that were indicative of the hysterical nature of their illness. For much of his career Babinski had felt himself under the sway of Charcot. On Charcot's death in 1893 Babinski felt free to develop his own ideas and in 1901 at the Paris Société de Neurologie he presented his paper "Définition de l'hystérie" in which he discredited many of the notions of Charcot and his school.

Babinski rejected many of Charcot's concepts of hysteria, the notion that seizures were an important part of the diagnosis and indeed that the condition was hereditary. Babinski built instead on Charcot's belief that whether a patient was hysterical or not could be determined through the neurological examination. He gradually realised that many of the "signs" of hysteria present in the neurological examination were also present in classical neurological disease such as myopathy. In this neuromuscular disease, in which the muscle fibers do not function as they should, a patient could unwittingly present as suffering from hysterical neurosis. Without tests for myopathy the conclusion could be none other than that they were neurotic and their inability to walk an elaboration of their own ideas and inward beliefs.

Babinski thus sought out a new direction as an adjunct to physical examination. He classified hysteria based on ideas of suggestibility and reversibility. If a patient could be cured by a good talking to then that was a sure sign they were hysterical. The problem was that despite being shouted at, ridiculed or sympathetically told their problems were emotional not physical, few patients actually got better. The neurological examination thus remained as a gold standard for determining whether a person's physical complaints were caused by the death of a relative, an incident reminding them of a difficult childhood event or other psychological trauma. It was this neurological examination that Hoover hoped to refine.

The problem is that even today, Hoover's sign still has only a limited evidence base[10]. It is not known how many healthy individuals are walking around without any awareness of illness, that if they were grabbed from the street they also might demonstrate this sign of hysteria. Lifting your leg on a bed is not something most people do every day nor can claim to have mastered any proficiency in. False positives can occur if a patient actively tries to demonstrate their illness in the faux environment of the consulting room. A patient can be diagnosed with hysteria through a rational desire to convince an otherwise disbelieving doctor. It also must be remembered that, "Hoover's sign is inaccurate if the weakness is mild."[11]

In antiquity hysteria had been considered a disorder of women and the womb. A woman's reproductive organs and their movement within the body were seen as the cause of a variety of symptoms. With their own limited knowledge hysteria became a repository for the Greeks of many "medically unexplained symptoms." In an age where female menstruation and sexuality was not properly understood, hysteria focused around false pregnancies, pre-menstrual tension and other feminine somatic symptoms.[12] In other words it was a vessel for any condition that medicine could not currently explain. Ironically the term "medically unexplained" persists today as a euphemism for hysteria, neurosis and the neurotic patient. In recognizing that scientific knowledge is a body of information prone to cultural as much as evidence based revision and that this "evidence" is interpreted in the light of dominant ideologies and belief systems, medicine has thus apparently failed to make any progress since Greek times. The understanding of disorders affecting the function rather than the form of the body as psychogenic is viewed without question as the correct one and we may laugh at the Greeks for their odd womb theories.

In terms of gnosology, the understanding of another's illness at first hand, if the current dominant cultural force within medicine cannot explain a patient's claim to be sick, then that patient often finds themselves viewed with suspicion. There is no technological assault on the root cause of their symptoms or if there is, it is curtailed by the physician's irrational belief that patients might think themselves more ill than they already are. They are labeled as suffering a "neurological problem" where the use of "" indicates the physicians skeptical attitude to the reality of their symptoms.[13]

The ancient Egyptians ascribed nearly all pain in a woman to her uterus; "if her eyes ache" it is "the fall of the womb in her eyes" if her feet bother her, it is also "the fall of her womb"[14]. Hippocratic physicians saw women who were not getting enough sex as particularly prone to hysterical symptoms. Sex was thought to moisten the womb and thus put an end to its troublesome wanderings, to stick it in place as it were. This line of enquiry was continued under Freud and still forms a part of most psychiatric interviews. Such discussions are of course merely culturally determined; they are also less expensive than any technological explanation. We may include in this the use of emerging technologies such as functional MRI, A measure of blood flow in the brain and also QEEG, a type of computerized EEG, a measure of electrical activity in the brain.

"No neurologist in the UK uses QEEG," states one Scottish based neurologist who shall remain nameless.[15] To which one may of course wryly reflect on the words of the 19th Century Northern Irish physicist and engineer, Lord Kelvin, that "X-rays are a hoax" and ask in the absence of space occupying lesions whether pathology is not any biochemical abnormality that can be measured and compared with "normal" controls.

Rather than embrace technological change and the ability to measure non-lesion based biochemical abnormalities, those physicians who promote the diagnosis of hysteria have repeatedly placed themselves within a quite different tradition. Like those who demanded Galileo to recant, their science is ultimately based on belief and dogma rather than molecular science. This may well be a dualistic approach to human experience. But, to define the experience of a sunny day in a series of chemical reactions is the duty of science, to describe and embrace the experience of a sunny day is the duty of art. Most importantly, science has the capacity through definition to redefine how we interpret and describe that experience. It would seem that only in hysteria has art defined science. Freud remains a great novelist and a poor scientist.[16]

Galileo attracted scorn as his observations via an instrument contradicted observations of the naked eye[17]. He was thus forced to defend a technique that we now take for granted but one that in the cultural milieu of the day seemed counter-intuitive and against church teaching. Similarly the father of QEEG Prof. Frank Duffy, a senior neurologist and lecturer at Harvard University, must apparently defend the notion that he is measuring biological and electrical abnormalities rather than magical ephemera. However those who promote hysteria as a psychosomatic diagnosis would indeed argue that what is being measured is caused by abnormal beliefs, that the very ideas a person holds have in fact somehow fundamentally altered their biological make-up. This is despite the fact that no fMRI (functional MRI, a measure of blood flow in the brain) study to date has ever shown the idea in a person's brain which might be causing them to be ill. As the neuropsychiatrist Dr. Jay Goldstein comments in his work "Betrayal by the Brain", such an approach to neuropsychiatry is a singularly British affair:

> *If the only tool you have is a hammer, everything looks like a nail." This aphorism is appropriate to the psychologising of neurosomatic [functional] disorders, especially by the British.*[18]

It therefore matters little what outward observations are made, the interest is not in disease or disease by which a person might think themselves diseased, but the psychological, social and cultural motivations that might make a patient think they are sick.

Prior to Hoover's time hysteria had been more associated in men with hypochondria, a condition where a person might be morbidly worried about their health. With Hoover's sign it became associated with deviance on the part of the individual. The neurological examination in its infancy thus became embroiled in concepts of how much people had willed themselves to be diseased or were willing others to think they were ill. No attempt was made to distinguish between feigned illness and illness a person experienced to be as real as the neurologist examining them. An illness that comprehensively convinced both the patient and those not part of the neurological and psychiatric clique that they, as a sick person, were suffering from disease. With Hoover's sign patients who sought to deceive and who were deceived could be detected by one and the same method.

A positive Hoover's sign can only ever indicate that the basic wiring of this portion of the central nervous system is intact. The fatal leap that can follow is when physical examination is used to dive into the murky depths of causation. To say that a positive Hoover's sign demonstrates that the patient is depressed or has converted emotions into a specific physical symptom, for example, is plainly absurd. Other tests, such as collapsing weakness are also associated with hysteria but a third of neurological patients, mostly those who have suffered a stroke, will also demonstrate this sign.[19] The doctor will therefore also make a value judgement as to the actual cause based on the personal history of an individual. The fact that this is little more than a value judgement and not evidence based medicine is seldom recognized. However, the English psychiatrist Eliot Slater stated:

> *Unfortunately we have to recognise that trouble, discord, anxiety and frustration are so prevalent at all stages of life that their mere occurrence near to the time of onset of an illness does not mean very much.*[20]

If life events were the cause of serious and long term disabling illness then the psychiatric profession would do well to extend its care to that of cancer patients, baldness, victims of heart attack, stroke and even stomach ulcers. The latter, a once clear sign of "stress" is now understood to be caused by bacterial infection. German scientists discovered the bacteria Helicobacter Pylori in 1875. It was quickly associated with ulcers though only in the late 20th century was this widely accepted. This basic medical fact still eludes some physicians. What then to say to patients with medically unexplained symptoms such as weakness? Dr. Chris Bass has the answer:

> *Stress is a common problem and can lead to hypertension and duodenal ulcers as well as what we call functional weakness.* [21]

This is not to say that stress is not a component in illness. Stress for example has a well-researched impact on the immune system.[22] RATHER IT IS TO ACKNOWLEDGE THAT WITHOUT UNDERLYING BIOLOGICAL FACTORS THERE WOULD BE NO ILLNESS. The student studying for exams may well develop "functional weakness" they may equally catch the flu. The association between stress and disease can be seen to be so much a part of all illness that it is useless as a diagnostic tool. Tests on laboratory animals have shown stress to have an impact on cancer growth and tumor type, but not as an actual cause of cancer.[23] In patients with "functional weakness" where changes in muscle tone do occur,[24] ascribing stress as a factor in the absence of any biological cause would be a unique moment in medicine. History shows us that illnesses once thought to be "hysterical" are, as science progresses, frequently reclassified as organic.[25]

Originally considered a psychogenic disorder, camptocormia, an abnormal posture with marked bending of the spine that goes away when the patient lies down, has become an increasingly recognized feature of parkinsonian and dystonic disorders.[26]

Though often very different in the way allegedly hysterical patients present to doctors the "positive signs" of hysteria are not fool proof and can lead to missed diagnoses. Studies that examine misdiagnosis rely on these errors being uncovered. In Britain there is only one neurologist per 177,000 people, the lowest in Europe. As we shall see, social and cultural pressures within medicine make the discovery of missed diagnoses unlikely.

Most neurologists comprehensively oppose re-investigation. This stems from a fear that such re-investigation will convince patients that they are actually ill and an institutional belief that they are not.[27]

We are therefore thrown back on what to say to patients whose conditions have not been explained by technological means. Current debates in neurology suggest the use of the term "functional" when discussing symptoms for which no organic basis has been found and in which signs such as Hoover's are present. Some neurologists view the use of the term "functional" as neutral. It is used to indicate that the condition may well have an organic basis but that psychological factors also play an important role. However the Oxford English Dictionary defines functional illness as:

> ***functional*** *(of a **disease**) affecting the operation rather than the structure of an organ: OED(2007)*

This is the way in which it was originally meant within the English language. The 17th century philosopher and psychologist David Hartley in his work "Observations of Man" repeatedly employs the word functional in its physiological sense, exactly how the Oxford English Dictionary defines the word. Medicine is not without tradition and Hartley uses "functional" in much the same manner as the 16th century English physician Edward Jorden in his work on hysteria, "The suffocation of the Mother" (Mother was a 16th century term for the womb). Both physicians use it as a term of reference to any poorly understood medical condition, though Jorden's experience of medically unexplained symptoms as proceeding from organic disorder is of particular interest.

In 1602 a 14 year old girl, Mary Glover, had been bewitched by "an old Charewoman", Elizabeth Jackson. As a result she had "fallen into fittes...so fearful that all around her supposed she would dye." Mary Glover became speechless and blind, "her neck and throat did swell extremely depriving her of speech" and later her arm and whole side were deprived of feeling and movement. To the physicians of the day such a presentation could not be explained. Yet there were clear physical signs, "her neck and throat did swell extremely depriving her of speech," that suggested a very organic nature to her illness other than simply globus hystericus, the subjective *feeling* of a lump in the throat. Indeed Jorden argued that her illness was due to natural causes rather than diabolical influence and witchcraft. Jorden lost the case and went away to write about his experience. Elizabeth Jackson who he had defended was sent to prison for a year. In his treatise written after the trial Jorden writes:
"the easie passage which [the wombe] has to [the braine, heart and

liver] by the Vaines, Arteries and Nerves...it is an affect of the Mother or wombe wherein the principal parts of the body by consent do suffer diversly accordaing to the diversitie of the causes and disease wherewith the matrix is offended," (Jorden 1603) pp. 1, 5

Some modern commentators such as Merskey[28] suggest Jorden is arguing that Glover's illness is due to perturbations of the mind. It is clear however that this is not the case. No doubt to argue that Glover's swollen throat and other outwardly visible signs were "all in the mind" would have led with the same reaction amongst the jury as is met by most physicians today who suggest that extreme physical and visible symptoms are psychogenic. Jorden would have found himself laughed at and ridiculed. In an age before the microscope the jury sought outward causation and magic was one socially accepted means by which the body might become physically and indeed perilously sick. That the mind of an individual might cause them to be ill or that the symptoms in some sense might be "unreal" or "non-organic" was beyond the reach of their social milieu.

What is clear in Jorden's work is that he is calling on the ancient doctrine of sympathy, "the principal parts of the body by consent do suffer diversely." He uses the concept that if one part of the body is hurt, another part may fall ill also. That "consent" is mediated now not through the movement of the womb but by the access the womb has through "Vaines, Arteries and Nerves." We thus see a primitive but very physical and organic expression of the notion that the body is an organism with interdependent parts and systems. Influenced by Hippocrates, Jorden places the root cause of Glover's illness with the uterus. Using the word matrix in its Middle English sense, as a synonym for "womb", he states that the various maladies of the body originate from the "diversitie of the causes and disease wherewith the matrix [womb] is offended."

Jorden loses the trial not because he proposes that Mary Glover is hysterical but rather because he rails against the notion that the sympathies to which hysteria was prone could be extended beyond the body. In effect he argues against the notion that spells, words and thus ideas could cause illness. He loses because he challenges the status quo in their concepts as to what might cause disease.

Building on Jorden's work Hartley developed his own theory in which he believed that functional disorders were caused by abnormal vibrations of the nerves. In this respect Hartley acknowledged he was influenced by Sir Issac Newton who believed the forces of gravity to be transmitted by vibrations in the Ether. In a throw back to Elizabethan notions of a universe in harmony, Hartley believed the human nervous system should also vibrate sympathetically with the cosmos.[29]

Another notable figure in the history of functional illness was Andrew Combe. This Scottish physiologist also had an interest in phrenology[a], the study of lumps and bumps on the skull. He wrote in his "Observations of mental derangement" in 1831 about the functions of the brain as suffering from illness whose "exciting causes may be divided into two great classes of local and functional." He suggested that the functional disorders were not only the "most

---

[a] Phrenology is a theory that claims to be able to determine character, personality traits and criminality on the basis of the shape of the head.

frequent but also the most important." J.Russell Reynolds, the prominent English Neurologist (1828-1896) contributed to the debate over functional illness in his "The diagnosis of diseases of the brain, spinal cord nerves and their appendages." Published in 1855 this work defined disease as organic and functional, the former term applying to observable lesions, the latter referring to physiological, unseen disturbances. His contemporary, Sir William Richard Gowers, suggested: "They are transient and not permanent and they are not known to depend on organic changes." Gowers did however suggest that there was an underlying physical cause: "Molecular changes in nutrition, considered as such, must be colossal to be detected. Such alterations, not sufficient to be seen but still considerable, probably constitute the morbid process in many diseases that are commonly classed as functional."

Gowers included chorea, paralysis agitans (Parkinson's disease), tetanus, migraine and hysteria within this classification.[30] Crucially it was recognized by the Yorkshire born neurologist Hughlings Jackson (1835-1911) that not all pathology was available to the scrutiny of the test or examining physician whilst the patient was alive. Some diseases could only be minutely observed once the patient was dead and through the discipline of careful post mortem.

Early in its history however the word "functional" was also used to indicate a psychological illness, a disorder of intellect or emotions. It was used as such by the German phrenologists Gall and Spurheim (*circa* 1800). They believed that by examining the shape of a patient's head they could uncover signs of the psychological affliction in the brain beneath. It was in this sense that Sigmund Freud in his later career began to use the word and this sense can also be found in current usage. The Collins English dictionary defines functional as:

> **Functional**: *Denoting a psychosis such as schizophrenia assumed not to have a direct organic cause: (Collins English Dictionary 2007)*

In contrast to Charcot's notion of these disorders as: "a cortical lesion but one that is purely dynamic or functional."[31] Freud asserted that:

> *The lesion in hysterical paralyses must be completely independent of the anatomy of the nervous system, since in it's paralyses and other manifestations hysteria behaves as though anatomy did not exist or as though it had no knowledge of it (Freud 1893)*[32]

This complex etymology makes the experience of patients with functional neurological symptoms confusing. This difficulty is increased as the symptoms they are experiencing have an awkward place in the history of medicine.

Hysteria, conversion disorder, functional weakness, functional neurological deficits are often seen as one and the same thing. In a view that largely arises from Freud's break with Charcot but as we have seen also has earlier precedents, they are seen as a weakness in the individual, an inability to cope with psychological or socially stressful life events. It may be used as an *aetiologically* neutral term by one neurologist but may well mean something very different to another doctor, who may well assume that the patient has been given a psychiatric diagnosis.

The situation is further confused by the fact that those who propose the use of the word "functional" as neutral fail to relate it to the often dire circumstances that patients find themselves in. The patient who is wheelchair bound or hemiplegic is unlikely to respond well to being told they have "functional symptoms". If they leave the consulting room punching the air with the words "Yes a diagnosis at last!" The clinician may conclude that the reason "functional" is acceptable is because they have no understanding of the etymology of the diagnosis. What patients are inevitably more concerned with are the *outward signs*, the tests and recognition from the physician that their illness has a medical reality beyond the confines of their own mind, this being the very essence of the condition they actually experience.

For the patient this is very much a disease, a process that has overtaken them and, as studies show, may well leave them severely distressed and disabled in the long term. To say that a wheelchair bound patient is suffering "functional symptoms" downplays the reality of outward **signs**, the very real fact that the patient cannot move their legs. The doctor may well be able to move their legs for them, or elicit intact reflexes with a well placed hammer blow, but the fact remains that the reason the patient is laid on the bed before them is that they themselves cannot move their legs. If they could, they would not be there. Having had it demonstrated that there is nothing wrong with them the patient has little choice but recourse to the theatrics of illness. Wheelchair use, complaints of pain and even their stubborn long-term refusal to walk, all become seen as abnormal illness behaviour given the lack of pathology. It is of course evident that illness is only perceived as theatre in the absence of discovered pathology, not whether pathology is actually present. As Gowers stated: "Such alterations, not sufficient to be seen but still considerable," could well constitute the disease the patient is struggling to explain and present.

Furthermore, as the neurologist Oliver Sacks recognises, this can be associated with flaccid muscle tone or other subtle changes in muscle function[33]. It is therefore hardly an accurate description of what, for the patient, is a very organic and distressing experience. It is a dangerous diagnosis as it assumes that the neurologist has a privileged understanding of the patient's own conscious awareness of their suffering and that the doctor's own knowledge of disease is total. The patient becomes like an astronaut telling a liberal flat earther that the world is round. The neurologist may well defend the patient's right to their "world view" but in reality very few neurologists actually believe it.

> *In clinical practice it is often difficult for a physician, faced with a patient in a hospital bed unable to use his or her legs despite normal tests and clinical findings, to differentiate between conversion disorder, factitious or fabricated disorder or frank malingering. What the clinician is being asked to do is to determine whether or not the symptoms are being produced intentionally or not; and what the motives are: Dr. Chris Bass (2006)* [3]

Bass is also on record as insinuating that those with cognitive deficits in Myalgic Encephalomyelitis (ME) are malingerers, stating that patients with ME fail tests for benefits that others with serious penetrating brain injuries requiring 24hr care would be able to pass.[34] This is despite the fact that the work of Tanaka and his team has linked reduced cognitive responsiveness in ME to a disorder of vasomotor function and reduced cerebral glucose metabolism.[35] Rather than blame his failings as a physician and the technology at his disposal, Bass blames the patient.

Historically medicine and religion are not diametric opposites and in primitive societies the roles of priest and healer were often blurred. Psychoanalysis had its origins in the confessional and the Judaeo-Christian tradition of illness has a moral basis and spiritual dimension. In treating patients with medically unexplained symptoms as social and moral deviants, somatisers and neurotics, contemporary medicine has regressed from its scientific roles and ambitions to that of asserting a moral and authoritative creed as to what is actually wrong with these patients, that this is "wholly psychological" and has "no organic basis whatsoever" (Consultant Neurologist, RVI Newcastle[36]). The breakdown in doctor/patient relationships that follows is an inevitable consequence of telling someone who experiences a physical condition that it is, in effect, "all in the mind."

Efforts to understand these conditions as functional disorders of the nervous system may well aim to prevent this breakdown but they are not removed from the historical tradition of referring any little understood condition to psychiatry. Descriptions in patient advice leaflets[37] that the patient is suffering a "software error" are understood in wider medical circles, amongst nursing staff and General Practioners as indicating that as the patient's central nervous system is intact then they are indeed either malingering or mad. The question must be also raised as to whether such explanations are good science. As Francis Crick, the pioneer researcher of human DNA has commented on computer models of the brain, "These people are good engineers, but what they are doing is terrible science! These people willfully turn their backs on what we already know about how neurons interact, so their models are utterly useless as models of brain function."[38] Likewise, computer metaphors of illness are utterly useless models of brain dysfunction and ignore the organic chemistry that underlies all illness.

2) Prognosis, Diagnosis & Treatment

It is ancient Egypt and a young woman is crouched naked over the smouldering dung of the crocodile. Her illness is associated with the god Sobek and as his symbol is the crocodile, it is through sympathetic magic that she is to be cured. The smell of the dung will match the smell from her womb and draw it back into place. Her priest, who is also her doctor, has prescribed her this treatment because it is written in the sacred Kahun Papyrus, the diagnostic manual of medicine in the old kingdom. As a young priest he studied under many great men, all of whom believed in the power of rank crocodile dung to draw the womb back into its proper place and stop it afflicting the body with weakness and other suffering.

Now our priest was not entirely convinced by this treatment, some of his patients got burnt in their nether regions, some went off to seek the advice of quacks and other "alternative medicine" types. The wealthy invariably asked for a second opinion that usually involved wine enemas or other trendy treatments. Nonetheless he was a career man and felt that his progression, if not his livelihood, depended on him keeping his mouth shut lest he be judged a maverick or a cowboy. The disease that he is treating is of course hysteria and though the Greek Physician Hippocrates also asserted that it was a disease of the womb and thus primarily a disease afflicting women, such associations became increasingly abandoned by the likes of 18th century doctors such as Sydenham and Whytt.

In the 18th century medicine began to move away from the idea that illness was caused by an imbalance of the humours in a person's body. The choleric or melancholic personality types of antiquity were seen as rooted in physical imbalances that could also give rise to classical disease entities such as cancer or gout. Hysteria did not immediately lose its associations with the womb. The condition in men continued to be referred to as hypochondria and increasingly became seen as a distinct disease entity. Hysteria retained its ability to imitate other diseases and Whytt described it as responsible for everything from cataplexy[b] to tetanus, vomiting of black matter and, in an age where good hydration was not understood, "A sudden and abundant flow of clear, pale urine."[39]

It would be a mistake however to assume that such "imitations" were all dropped from the diagnosis of hysteria as medicine developed in the 19th and 20th centuries. However even in the 18th century there lurked the suspicion that hysteria was indeed a diagnosis of "ignorance and error". The physician Thomas Willis (1621-1675), who spent considerable time studying hysteria, writes:

> *When at any time a sickness happens in a woman's body, of an unusual manner, or more occult original, so that its causes lie hid, and a curatory indication is altogether uncertain....we declare it to be something hysterical...which oftentimes is only the subterfuge of ignorance.*[40]

---

[b] Cataplexy is a sudden attack of muscle weakness, it is associated with narcolepsy, a disorder whose main signs are excessive daytime sleepiness, sleep attacks and hallucinations on waking. It is classed as a neurological rather than a psychiatric condition.

Willis recognises that hysteria is in fact a disease projected on to the patient by the ignorance of the physician. In contrast to Freud's later ideas he asserts that medically unexplained symptoms do not arise from psychological pathology. Willis does not believe that there is no organic basis behind medically unexplained conditions, he asserts that the term "hysteria" is as a catch all diagnosis for any medical curiosity that cannot be explained by conventional means.

Most strikingly, Willis identifies the diagnosis of hysteria as associated with "unusual" symptoms, so that at the very beginning of pre-modern studies of hysteria, the idea of any symptom that defied current medical thinking and contemporary scientific understanding (and this includes inconsistency of presentation) was seen as a key diagnostic aspect. The patient who can move their arm perfectly well when distracted is unusual as they are inconsistent and therefore they have already demonstrated a "sign" of hysteria.

The lack of any medical explanation for these curios resulted in an increasing reluctance amongst some physicians to see hysteria as having an organic basis. Joseph Raulin, for example, believed that it was a disease "in which women invent and exaggerate and repeat all the various absurdities of which a disordered imagination is capable," and that it, "has sometimes become epidemic and contagious."[41]

This was not yet a universal opinion amongst the medical profession. In one interesting insight and in an age before functional MRI, a measure of blood flow in the body and brain, before tilt table testing[c]

---

[c] Tilt table testing is a technique in which heart rate and blood flow are measured when the patient is lying down and then when the patient is tilted to an upright

and before complex blood analysis, Georg Ernst Stahl blamed an "increasing heaviness" of the blood and disorders of the circulation as the root cause of hysterical symptoms. Foucault comments:

> *Georg Ernst Stahl opts instead for an increasing heaviness of the blood, which becomes so abundant and so thick that it is no longer capable of circulating regularly through the portal vein; it has a tendency to stagnate, to collect there, and the crisis [or seizure] is a result 'of the effort it makes to effect an issue either by the higher or the lower parts.* [42]

Like Stahl, Charcot defined seizures as an essential part of the diagnosis of hysteria. We now know that disturbances of the circulation and the heart can cause venous pooling, clotting, cardiogenic seizures and syncope, just as much as abnormal brain activity.[43]

Antiquity had defined hysteria as a disorder of the womb and with the increasing realisation that men also suffered the same malady, a mechanism was required to explain how, in the absence of that particular reproductive organ, men also could be described as "hysterics". Sydenham and his contemporaries began to see hysteria as a disorder of spaces within the body. If the womb was the most hollow organ in the body then surely any organ that could be easily occupied, such as the stomach or the larynx, could harbour the

---

position. It can be used to diagnose a number of pathologies including Postural Orthostatic Tachycardia and cardiogenic seizures. Both are conditions related to poor venous return and an increased risk of clotting.

origins of hysteria in the susceptible patient? "Susceptible" and "occupied" suggests merely that their organs were more easily penetrated and controlled by "animal spirits" from the brain, the person's baser nature. Hysteria therefore became associated with deviance rather than disease.

Women remained more prone to hysteria due to the simple fact that they had more "space" in their body, a body in which the fibres were less tightly woven together than in the muscular male. The assumption was even made that women who laboured on the land would be less prone to the disease than the more delicate emerging middle classes.

This ability to resist penetration was for Sydenham no less than the ability of the soul to resist corruption and keep its thoughts and desires in order. Sydenham, as Foucault makes clear, was not just concerned with the physical body in which the disease was observed by the physician, but the metaphysical body in which the disease (or the idea of disease) had taken root. The link between these remained a problem; it was resolved by reference to the action of the nerves. The nerve fibres of the brain and central nervous system were seen as having an exceptional sympathy with this non-corporeal body. Essentially, they were able to vibrate in harmony and therefore convey the symptoms of hysteria as physical afflictions to the body of the patient. They were able to convert emotional discord into physical symptoms. That Stahl's rather fanciful explanation of poor circulation should give way to ideas ultimately rooted in renaissance notions of the "harmony of the spheres" and the patient's body in tune with the cosmos is interesting. It speaks volumes for medicine's tendency to embrace the fashionable and culturally moulded explanations of disease rather than objectively study affliction and the afflicted.

This idea of sympathy was not an isolated relationship between the body and any metaphysical entity. Whytt, for example, viewed the body as having a sympathetic tendency within itself, so that an affliction in one part of the body could spread and be mediated to another part by an over sympathy within the nervous system. In time the notion of sympathetic action gave way to the idea of irritation:

> *The list of possible "causes" of hysteria were absurdly disparate: nervous shocks, pneumonia, exaggerated religious practices, membership of certain professions, races, and religions, and perhaps most tellingly, a variety of sexual behaviors, including masturbation, "venereal excesses," and sexual diseases. (Gantz 2005)*[44]

Above all, hysteria acquired a moral dimension in which it was the indulgences of the individual that had "irritated" their nerves and thus produced the signs and symptoms of hysteria. This was not however a literal inflammation, but rather a hidden and essentially meaningless reference to the notion that the nervous system was out of sorts, out of harmony, with no organic cause and thus in reality the concepts that 19th Century physicians inherited as "nerve doctors" were not entirely removed from their 17th century supernatural origins.

It was in this world that Charcot began to develop his ideas of hysteria. Writing about hysteria in 1889 he said:

> *There is without doubt a lesion of the nervous centres but where is it situated and what is its nature? ... Certainly it is not of the nature of a circumscribed organic lesion of a destructive nature ... one of those lesions which escape our present means of anatomical investigation, and which for want of a better term, we designate dynamic or functional lesion.*[45]

This notion of functional lesions of the nervous system was essentially a development of previous notions of sympathy. Given the history of the term and the battle between psychiatry and physiology for ownership, we cannot pass over Charcot's own frustration with the inadequacy of the word "for want of a better term." At first Charcot assumed that these lesions must also have an organic cause and that the microscope would reveal the nature of this sympathy. In an age without MRI or CT or even QEEG scans[d], Charcot allowed himself to be influenced by others towards a non-organic explanation of any symptoms he was unable to explain by conventional means. Even in our own time it is claimed by some that QEEG[46] is set to revolutionise the detection of brain dysfunction and further assist diagnosis.[47]

---

[d] MRI and CT are now commonplace means of locating physical lesions, QEEG is a computer driven type of EEG a means by which the electrical activity in the brain can be measured. It has a "higher resolution" than standard EEG. The author was party to a QEEG study performed by Frank Duffy MD the "father of QEEG." This was of a patient suffering repeated closed head injury. It can be found in the notes at 24, permission is granted for further distribution.

Charcot was impressed by the work of the English doctor J. Russell Reynolds. Like Reynolds he argued that a patient could become immobile by simply thinking about being paralysed. He classified hysteria as a neurosis together with epilepsy, chorea and parkinsonism[48] in contrast to structural lesions. Charcot stated that the idea of paralyses arising from thoughts, "had long been known but only Reynolds had studied them in a methodical and systematic way."[49] Alongside emotional disturbances, Charcot also believed that hysteria could be caused by prolonged exposure to a damp cold climate. Here he was inspired by a case described by Moritz Romberg (after whom the Romberg test is named), in which a man developed a tremor having been robbed by cossacks and left in the snow[50]. Deciding between emotional and physical trauma was not an easy task. In some circumstances Charcot was clearly prepared to hedge his bets over the aetiology of both hysteria and the illness patients suffered.

Charcot suggested that patients who had inconsistent weakness, or paralysis that fluctuated or did not conform to anatomy, were suffering from these dynamic or "functional" lesions. Charcot's contemporaries saw reversibility of symptoms and inconsistency in patients as supporting evidence for the view that there was no underlying pathology. Though a master of dissection, Charcot was limited in his ability to understand neurology on a cellular level. Thus diseases that are no longer viewed as essentially psychological, such as Parkinson's, were all placed in Charcot's basket of "functional" and by his later definition, neurotic illness. The electron microscope remained a wonderful invention of the future, as indeed did many of the staining techniques commonly used today with conventional microscopes.

Ultimately Freud, who believed hysteria to be a neurosis without any organic basis, influenced Charcot into abandoning his quest for an organic mechanism. This was the first disaster for patients deemed to have medically unexplained symptoms. Even worse, in Freud's bid for dominance within psychiatry the idea that emotional disturbances could be the sole cause of physical symptoms not only became entrenched within psychiatry but at the very core of medical thinking. This meant that when met with symptoms that could not be explained in conventional biomedical terms of the day, clinical practice would draw a line under organic investigation and pass the patient on to psychiatry.

In the 21st Century should patients with medically unexplained symptoms be passed on to a psychiatrist? At the same time as claiming neutrality in the debate the Edinburgh neurologist Jon Stone and his colleague the psychiatrist Michael Sharpe repeatedly say "Yes."

> *A patient who is unimproved after receiving a careful explanation, a trial of antidepressants and physiotherapy should probably be referred. - (Stone et al 2005)*[51]

So that Freudian notions of medically unexplained symptoms have not gone away, anything that is not understood is seen as hysterical neurosis. The neurologist Oliver Sacks includes some types of migraine within this category[52]. If the patient gets better, they have recovered from their neurosis, not from missed pathology. The explanation they receive must not be an affirmation that they are suffering physical illness. Whilst pharmacological treatment in the form of anti-depressants reinforces the notion that hysteria is an active process on the part of the patient. Their will to act is depressed and must be lifted out of the mire.

The patient themselves is often seen as refusing or lacking the ability to come to terms with the heterogenous[e] view that there is nothing physically wrong with them, essentially a character flaw. No suggestion is made of using drugs that as either inhibitors or exciters are not part of the panoply of psychiatric medicine[f]. If symptoms persist it is not because of the severity of missed or little understood organic pathology but the fault of the neurotic self.

Diagnosis and treatment stems from the psychiatrist's couch rather than the benchmarks of science, the electron microscope and the laboratory.

---

[e] In other words the imposition of an external view and interpretation on that of the patient, even though the physician cannot experience what the patient is suffering, nor often can the patient articulate it.

[f] On taking 30mg of Baclofen, a drug prescribed for *neurological* symptoms the author, alleged to be suffering from "hysteria", found symptoms of spasticity massively improved so that he was able to do star jumps. The neurologist stated in writing that he had no idea why this might be the case. Collapses accompanying the star jumps did not instigate any vascular investigations for over 3 years.

3) A Brave New World

There will come a time when our descendants will be amazed that we did not know things that are so plain to them-
   Seneca (first Century) Natural Questions, Book Seven

All diseases of Christians are to be ascribed to demons-
                                                St. Augustine

In the early 20th Century the diagnostic criteria for hysteria shifted again. Following on from Babinski, the neurologist Arthur Hurst suggested that a key aspect to the diagnosis of hysteria was in fact outcome, the notion that symptoms could be easily reversed. Hurst treated the victims of shell shock at Seale Hayne near Newton Abbot in Devon, England. At a time when there was little sympathy for victims of shell shock and treatments ranged from electric shock therapy to solitary confinement, Hurst took a humanitarian approach. This approach continues to echo through history, an editorial on 16th August 2006 in the Daily Telegraph said of men suffering shell shock and shot for desertion: "These were not rational deserters but men driven mad by suffering. The pardons reflect a welcome change to our attitudes to life and death in battle."

It would seem apparent that in the near 100 years since The First World War we have made steady progress to an understanding of neurosis and medically unexplained illness as having their basis in the brain and not purely in the individual's social and cultural context. That in contemporary society we can acknowledge depression and other disturbances of brain function *as disease*. Indeed not to do so is no longer culturally acceptable or beneficial to patient relationships.

Yet such a progression is not universally acknowledged, for example the 21st century psychiatrist Simon Wessely, whilst acknowledging the biological correlates of functional illness, has attempted to shift the blame back to the patient for the perpetuation of their illness and ownership of the concept of disease back to the physician.

Wessely asserts that the very label of illness prevents patients from getting better:

> *By 1916 it was accepted that many men could break down if pushed long and hard enough. But if a person was fundamentally 'sound', provided that he was managed correctly—and, in particular, not given a medical label nor sent to a rear hospital for a prolonged period of time—this condition ought to be short lived.*[53]

This is a common theme in Wessely's work, rather than giving the sick in society the power to name, own and overcome their condition, Wessely sees such powers as the explicit right of the physician. Patients require *management,* not treatment. This even applies to those suffering "functional" illness as a result of war.

In the case of a first world war soldier who was shot at dawn, a Private Harry Farr, Wessely retrospectively diagnoses him with an anxiety disorder, a phobia of war. In writing about Farr's case, whilst decrying the death penalty for the horror it is, he acknowledges the need for military discipline in difficult circumstances. He quotes with approval the military historian Richard Holmes: "It was indeed a hard law but it was, in general, fairly applied."[54.]

Farr's failing, as Wessely suggests, is not of a rational man driven mad by suffering but of a man judged by his peers as deviant enough to display his suffering via illness behaviour and a refusal to fight. Wessely emphasises against a canvas, in which real men repressed their terror, that it was before a major engagement Farr chose to break down. He writes:

> *And that night Farr would not have been alone in experiencing intense fear— there were probably few around him who did not feel something similar as they faced the prospect of attacking the notorious Quadrilateral the following morning. What the Court Martial had to consider was that Farr did not control his fears, whilst his comrades did.*[55]

It remains the case though that in May 1915 Farr's position had been repeatedly shelled. Some critics of Wessely's suggestion that it was pure psychology that led Farr to break down suggest that Farr suffered damage to olivocochlear bundle in the inner ear[56]. This would have destroyed the ear's ability to filter sound making any noise unbearably painful.

Unless we assume that the German gunners were so inept as to be unable to hit their target at all, Farr must have been near enough for acoustic shock. Concussion that would cause not just the psychological insult of fearing for his life but the mild traumatic brain injury that is increasingly associated with the concussive effects of high explosives and related anxiety and depression in its aftermath.[57]

With the limited investigative techniques of Hurst's time it is unlikely he would have found micro-hemorrhaging in the brain[58] whilst Farr or any other shell shock victim was alive. Yet as Webster comments:

> *One man had been in a dug-out when a shell exploded ten feet away. He then suffered tremors, general depression, and periods of crying. The next day he was unable to talk. That evening, he entered "a state of acute mania," shouting "Keep them back, keep them back." He was "quite uncontrollable and...impossible to examine. He was quieted with morphine and chloroform and got better and slept all night....Next morning he woke up apparently well and suddenly died." The second soldier was in an ammunition shed when it was hit by a shell. He became unconscious and died soon afterward. When the brains of these two soldiers were subjected to post mortem examination, microscopic haemorrhages and other vascular changes were observed. It was eventually acknowledged that the "overpressure effect" caused by a shell's explosion could rupture the internal organs of a human body and cause fatal injury, yet to do so without leaving any external marks.[59]*

Hurst's assumption was much the same as Charcot's: that visible damage to the actual tissue of the nervous system would mean that the patient was left with permanent neurological damage, this was in sharp contrast to functional disorders, which Hurst associated with abnormal function rather than anatomical damage cause by for example, high explosives.

Concussive injuries aside, in the extraordinary circumstances of the First World War, Hurst was faced with many patients wounded in the trenches whose symptoms seemingly outweighed the injuries they had received.

Treating a soldier, who had lost a thumb but developed paralysis in the arm, Hurst suggested that the paralysis was caused by recoil within the central nervous system from the site of injury. Faced with a soldier blinded by mustard gas who had "organically" recovered but who was still blind, Hurst suggested that the soldier had become so used to not seeing that they needed to learn how to see again. Hurst had little understanding of plasticity within the cortex and the ability of the brain to "re-wire" itself. Nor did he have any understanding of the complex haemodynamic response necessary for brain function, the association between blood flow in the brain and the necessity of this for neural processes to be activated.

Influenced by Janet, Hurst used contemporary ideas of "suggestion" to further explain the ways in which soldiers succumbed to "hysterical injury." That, for example, if a soldier suffering concussion were told by an authoritative figure, such as a doctor, that he might not recover, then his chances would be greatly diminished. Hurst used counter suggestion and his own status and authority to treat patients. That he was able to do so speaks volumes for the general loss of authority medicine and other professions, such as teaching, have suffered within society. We no longer live in a society governed by hierarchical diffidence.

The error in interpreting Hurst's work is to assume that there was no pathology behind the symptoms he faced, that the men were indeed possessed by ideas of illness. The critical fact is that the majority if not all of Hurst's cases were associated with a physical rather than an emotional trigger. This is demonstrated by Hurst's approach to his patients. Treating a boy who had suffered a head injury and subsequently become deaf Hurst gradually reversed years of disability by encouraging the boy to pay attention to the process of hearing once more, in essence a kind of internalized physiotherapy, an active fine-tuning of connections in the brain desynchronized by earlier injury.

In neurology today there is an almost neurotic fear that patients will feel that they are powerless victims to their illness and that if they believe their nervous system to be damaged they will not get better. This often runs counter to the patient's expectations of the disease process. Neurology's insistence that patients suffering medically unexplained symptoms are not suffering "disease" alienates patients from doctors who effectively deny them the use of that very noun which sums up their condition.

It is also a belief that runs counter to those professionals involved in the active rehabilitation of patients with neurological symptoms, explained or otherwise. As one physiotherapist trained in neurological conditions states, "If the central nervous system was unable to recover from illness, we would be out of a job."[60] Indeed this is the bedrock of rehabilitation for patients with neurological disease. The neurologist Ramachadran and his colleagues have studied instances in which an amputee still feels as though their arm is attached (this is called phantom limb syndrome.) Due to the layout of the motor cortex in the brain with, for example, areas that deal with the hand positioned next to those that deal with the face, patients often feel that if their face is touched then their amputated hand is being touched also.

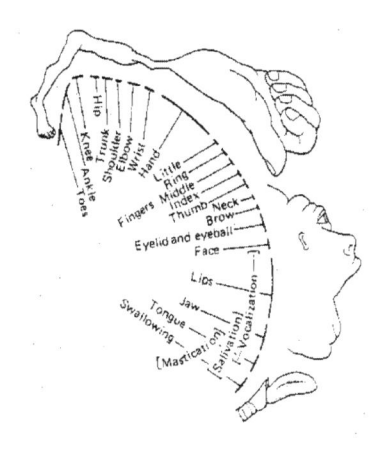

Figure 0: *Map of the Motor Cortex. Here we see how, for example, the thumb is adjacent to the neck. Brain dysfunction can cause motor circuits to colonise adjacent areas. The symptoms produced would, in the past, be considered as inexplicable and often hysterical.*

Ramachandran suggests that there are pre-existing connections between adjacent areas of the motor cortex. When the nervous system is damaged or disrupted biochemical changes cause these connections to be turned on. In the case of amputated limbs this can result in "phantom" pain from the amputated limb. As Ramachandran's work eloquently demonstrates, symptoms that apparently defy reason can have an organic rather than a psychosomatic basis. Hysteria also need not be the emotional disturbance described by Sigmund Freud. It is of course an obvious step to suggest that medically unexplained symptoms have a range of causes and are not just represented by the work of Hurst and Ramachandran. The history of any symptoms that defy current medical thinking suggests, however, that explanations all too frequently rely on notions of childhood abuse, illness behaviour, stress or downright malingering.

Neurology's understanding of the brain had its birth in the dissections carried out by Charcot. It evolved as a discipline mainly concerned with diseases that caused anatomical lesions or abnormalities. Psychiatry came to deal with conditions that were seen as neuroses, disturbances of the self. As we have seen, Freud essentially brought about the split between the professions of psychiatry and neurology. Freud focused on his own ideas of psychoanalysis and the unconscious without using any scientific method to support his explanations of the symptoms he found in patients.

Some practitioners of modern neurology attempt to re-integrate the two disciplines in both the way they diagnoses patients and their approach to treatment. In reality however this third way does not exist within the NHS.

One experimental approach to treatment that took place in London in 2006 was to run a clinic for patients suffering "functional" movement disorders. It was a joint venture between the neurology department of King's College Hospital and the department of psychological medicine. It adopted a multidisciplinary approach involving physiotherapist, psychologist and neurologist. The experiment was not developed, as it was felt "not to be a cost-effective use of time." When the one patient who took part in the experiment failed to attend a follow up appointment the whole idea was shelved. The clinic claimed to maintain an aetiologically neutral stance as to what had actually caused the illness; this in effect ran counter to the tradition of medicine.

More so than any other science, when in reality the oceans of knowledge are boundless and uncharted, medicine has constantly claimed that it has drained them dry. Writing in the "Madness of King George" Bennett describes how the King's doctors become obsessed with the consistency of his stools, a sign they understand. Discoloured urine is ignored because they simply do not grasp the significance. In the NHS where patients must fit into diagnostic boxes, be managed by distinct disciplines, who will actually manage patients where the cause of their symptoms is unclear remains an issue.

It is widely acknowledged that patients with functional symptoms are seen as a management issue. Tradition in neurology dictates that since they have no "structural disease" their treatment should be passed on to a psychiatrist. This is in marked contrast to the work of Arthur Hurst, who saw the total care of patients with functional symptoms as his responsibility. This was different from many of his colleagues who still sought to define the vulnerable individual as one prone to hysteria and thus the preserve of psychiatry.

Today, as in Hurst's time, the majority of neurologists and psychiatrists faced with a patient who has no obvious organic trigger for their illness will then attempt to identify the cause within the personality of the individual. As the philosopher Foucault realized it is wrong to assume that a dialogue with psychiatry is a discourse with reason. Sexual abuse that the patient is unable to remember, abnormal illness behaviour (often in the face of disbelieving doctors as the psychiatrist Eliot Slater recognised), maladaptive coping strategies; all are seen as the weakness in the individual that has caused them to develop their hysterical illness. The term "functional" quickly loses its neutrality. These individuals are somatisers, people who, unable to cope, translate their emotional trauma into physical symptoms. The problem is as Per Dalen recognizes in his essay "Somatic medicine abuses psychiatry" that illnesses with bodily symptoms have little in common with mental illness:

> *Since I am a psychiatrist, I have for a long time been intrigued by the extraordinary use of psychiatric causal explanations for illnesses that not only go with predominantly somatic symptoms, but also lack any basic similarity to known mental disorders. Are patients being helped by this peculiar way of interpreting their illnesses? No, it would be a gross exaggeration to maintain this, at least when there are pronounced complaints of some considerable duration. - Per Dalen (2003)*[61]

Whilst other diseases of function such as migraine and Parkinson's have had substantial research and pharmaceutical input, the treatment and understanding of "functional symptoms" lags behind.

It remains in a Freudian hinterland where cause and effect are cobbled together no matter how outrageous or bizarre. Even when examination has failed to find any psychic trauma, neurology battles on regardless:

> *When I send a patient to a psychologist I'm told there is nothing wrong with them, but there is, that's why I send them to a psychiatrist he always finds something wrong:*
> *Consultant neurologist, RVI Newcastle-upon-Tyne*[62]

The assumption often seems to be that as diseases affecting the function of the brain are reversible they a) cannot reflect any underlying pathology b) that they are disorders of the self. Both are odd assertions, as odd perhaps as asserting that a radio with an intermittent fault that eventually rights itself is suffering from some kind of neurotic disturbance. Of even greater concern is neurology's fear that investigations of hysterical patients will actually find something, as Stone et al state:

> *The need to look for disease also needs to be balanced against the risk of uncovering laboratory or radiological abnormalities that have nothing to do with the symptoms but which may delay or disrupt positive management.*[63]

So the question must be asked as to what exact pathology are we trying not to uncover; do we mean cancer, multiple sclerosis even liver disease? In 2006 liver disease in the case of Primary Biliary Cirrhosis was found to be associated with autonomic dysfunction and impaired baroreflex sensitivity via lesions in the globus pallidus.[64] Laboratory abnormalities viewed as unrelated to symptoms by one generation of doctors may turn out to be key instruments in diagnosis and treatment by another.

It is in this darkness that we start to dimly see that the body is not a system of discrete units but one interdependent organism. To use a car analogy, it is not just an engine, gearbox and oil but rather an engine dependent on a working gearbox and the correct oil. Even if the engine is intact the wrong oil may cause the engine to fail.

In contrast to this approach, neurology often inhabits a surreal world in which treatment has less to do with exhaustive testing and more to do with odd psychologically based tactics. Cognitive behavioural therapy is one technique employed in treating hysterical symptoms but as one Channel 4 documentary "Hypochondriacs: I told you I was ill" highlighted, its actual use is often tantamount to an assault on an individual's misfortune to be sick.

Though obviously not a hypochondriac the documentary included the case of a 37-year-old woman called "Jane". Five years previously Jane had collapsed and has not walked since. She is in a wheelchair, her car has hand controls and she is clearly perturbed at the failure of medicine to find anything wrong with her. A review of the documentary in "The Guardian" newspaper reads, "These idiots think they've got cancer, and Aids; they believe they're having heart attacks, strokes, and that they're paralysed. They're totally fine, physically."[65] This article immediately raises concerns.

That Jane, who knows herself to be *physically* disabled, should find herself amongst those who fear that they might end up terminally ill or in a wheelchair themselves, clearly demonstrates the inability of the documentary makers and newspaper journalists to differentiate between somatoform disorders and possible missed diagnosis.

Startling though this error might be the notion that there is only one somatoform disorder and that DSM criteria (a manual for physicians to aid diagnosis of mental illness) are singularly unhelpful, lies behind much of the thinking behind the use of the term "functional symptoms." Advocates call for a return to a so called golden age before Freud in which doctors treated disorders of the nerves and neurasthenia[g] was the fashionable diagnosis for any medically unexplained condition.

That the journalist Nicholas Wollaston should refer in his Guardian article to Jane as an "idiot", speaks volumes of the tour de force that medicine has achieved in some circles of western culture. Whilst quantum physicists may throw their arms up in despair that a unifying theory of physics is yet to be realized, no doctor ever admits to a patient that they simply do not know what is wrong. That Jane could well have been a victim of this arrogance was not even given a moment's thought.

In the documentary, Jane, clearly deemed to be suffering from "hysterical" paralysis, was asked by her psychiatrist Dr. Lars Hansen working alongside Dr. Florian Ruths of the Maudsley, to identify those things of which she felt most afraid. Rather than catalogue a host of personal trauma preceding her illness, the concerns she wrote

---

[g] Neurasthenia was a disorder characterised by fatigue said to be caused by "weak nerves." George Beard coined the term in 1869.

on the balloons offered to her, all related to the medical profession. Having symbolically burst these balloons her psychiatrist cajoled her to dress up as a clown and, though clearly distressed, she was then asked to wheel herself through the high street. Needless to say she remained paralysed. Hansen seemed to be oblivious to the fact that the only reason she followed his therapy regime was the same motivation that prompts the chronically ill to seek help from crystal healing and other fringe therapies, sheer desperation. Jane's asides to the camera clearly showed that she was far from sold on his explanations and even less so on the logic behind Cognitive Behavioural Therapy (CBT). Though the term "hysterical woman" has become a term of abuse it's association is no means accidental. It was not until the 1950's that Alberto Marinacci observed that some hysterical cases "had a hysterectomy in 1938, in the hope that the mental symptoms would be relieved, but the procedure was not beneficial."[66]

CBT has a further dimension in that many of its proponents, Simon Wessely, Mike Sharpe and Chris Bass, have strong ties with the employment insurance industry. So the question must be asked whether the priority is returning a patient like Jane back to health or back to work. Ruth Harrison, an insurance nurse for Unum Insurance Limited, points out that cognitive behavioural therapy has its ultimate roots in psychoanalysis. Whilst analysis of dreams may not see a person returned to health, the moral nature of Freud's work cannot be avoided. Harrison states that when therapy fails it is the person who is to blame, "It could be that in these situations individuals, despite what they say, do not actually wish to change and are comfortable with their way of being."[67] To this extent they are as

Charcot suggested of some of his patients, disabled as a matter of choice.

CBT continues the trend that Foucault identified. In the past psychiatry attempted to control the body, to shut away those who offended science and reason within asylums. Today, functional symptoms are controlled via attempted re-programming of the individual and the ideas that possess them. *Who requires re-programming is determined by whether any pathology has been discovered, not by whether any pathology is actually present.* The more resistant to psychological approaches the individual may be, the more deviant they are seen to be:

> *"So in future," Dr. Piet said pointedly, as he drew the session to a close, let's not waste any more time. Let us have no more agonizing about unimportant details. Let's get right in there to the basic difficulties, which are the cause of all these symptoms. And do me a favour- Let's have no more histrionics about perfectly normal absent-mindedness. After all, Karen," he smiled to soften his bleak conclusions," it's not as if you ever did anything very dreadful at these times. Making a cup of coffee? Going to the library? Come on! You, of all people with your gothic imagination can do better than that! Let's see what you're really like when you lose control. Surprise me!"*
> -The Spiral Staircase, Karen Armstrong

Karen Armstrong was later to be diagnosed with temporal lobe epilepsy by the neurologist Dr. Wolfe. For three years, including one psychiatric admission, she had been wrongly diagnosed and treated. Wolfe described her as "a text book case." In terms of treatment, rather than raise suspicions that her symptoms might not be caused by an emotional disturbance, the lack of reversibility and the lack of improvement merely re-inforced psychiatry's notion that the problem lay with Karen herself. Such outrageous interventions are of course not isolated incidents within psychiatry. The Royal Victoria Infirmary in Newcastle-upon-Tyne routinely makes use of the services of a Dr. Douglas Turkington in treating patients with functional symptoms. Turkington is on record as having organized an exorcism on a schizophrenic patient,[68] a fact that emphasizes the similarities between psychiatry and religious office. Shakespeare writes:

> *If thou could'st, Doctor, cast*
> *The water of my land, find her disease,*
> *And purge it to a sound and pristine health*
> *I would applaud thee to the very echo*
> *That should applaud again*
> *Macbeth Act V Scene 3*

Turkington may not be on record as using a patient's urine as a diagnostic tool for uncovering the cause of functional symptoms, but a colleague's comment that Turkington "always finds something" when mere clinical psychologists fail, paints him more as the Witch Finder General than a serious scientist. It also highlights again, as Foucault recognized, that what is unknown within medicine is seen as a deviance on the part of the patient, not a failing on the part of the physician. That by labelling the patient as deviant in some way the moral and scientific authority of the physician is preserved. The extent to which hysteria is a deficit of the will is central to the debate, as the British pathologist and surgeon Sir James Paget remarked in the 19th Century "...they say 'I cannot'; it looks like 'I will not': but it is 'I cannot will'."

Darwin demonstrated that species with plasticity in their neural responses to environmental factors tended to be more successful compared to those limited to simple reflex impulses. Whilst Paget recognized a deficit of this ability in an individual to mediate their response to their environment, the legacy of Charcot and Freud has been to interpret this as a flaw in the individual. They do not move because they are emotionally disturbed and thus unable to will any movement. This failure to see concepts such as "will" and the "individual" within medicine as anything more than complex organic processes produced by evolution has caused research on hysteria to focus on social rather than scientific explanations.

The reversibility, under certain conditions, of hysterical symptoms is seen as further evidence that the flaw is with the person rather than of the complex interactions between organic circuits, the haemodynamics of blood flow in the brain and biochemistry that allow a person to remember how to move, or indeed stop moving.

Under sedation it is claimed that many of this type of patient effectively recant their illness and the confession in the form of a film is stored for later viewing. It is assumed that this footage shows evidence of non-organic illness and that there is no underlying pathology. Indeed one BBC documentary showed a girl with a movement disorder in her leg. Treatment involved confining the leg within a plaster cast. A course of treatment not merely symbolic in its confining and controlling of what medicine cannot explain.

Coming round from the anaesthetic and on the resumption of movement the girl was clearly in agony. She was however told that as the movement had ceased under anaesthetic, there was no pathology, the fault in essence was with her. Sleeping and awake states of consciousness naturally produce changes in muscle tone; this seemed to elude her physicians. Low muscle tone during REM sleep has been extensively studied.[69] At no point was a problem with neurotransmitters or any admission of ignorance entertained, the bizarre assumption was that as her condition had not yielded to standard tests, it was not and could not be organic.

Whilst Hurst sketched out his theories for the organic basis of functional symptoms most modern approaches use sedation to indicate precisely the reverse, that there is no organic basis. Yet again we must reflect that to move my arm I must consciously know that I can move my arm. If there is a fault with circuits that allow me to know or remember how to move then movement will fail even if the circuits that actually produce the movement are intact.

There need be no fault with the actual person themselves, no history of emotional trauma, no disturbed childhood, no events so repressed that only a Turkington can cast the waters and find them out. That psychiatry should assert there is a link stems from theories of somatisation and the way medicine views patients, a view very different from the way patients often view their doctors.

3) Alas, the Storm Is Come Again

Again and again the diagnosis of conversion disorder, functional weakness or the myriad of other names given to hysteria stems from an inaccurate knowledge of the history behind the diagnosis and the influence of Charcot and Freud. It is a lack of knowledge that few are able to recognise:

> *There is no generally accepted explanation for how a psychological stress can convert into (often highly selective) symptoms. In this respect conversion hysteria retains "the doubtful distinction among psychiatric diagnoses of still invoking Freudian mechanisms as an explanation."..The crux of the problem is to explain how abnormal psychological states can produce specific, long term neurological symptoms and disability in patients (who claim not to be consciously responsible) in the absence of detectable pathology*[70]- Dr.Peter Halligan (2000)

The crux of the matter is perhaps not how abnormal psychological states can produce specific symptoms but as to why there should be an assumption that such symptoms are psychological in the first place. It is often the case that individuals present to doctors with physical symptoms that are interpreted via a leap into the diagnostic darkness, an assumption that the cause must be emotional. Just because no detectable pathology can be found does not mean that there is none. A fisherman who returns home empty handed does not claim that the sea is empty.

Patients with functional weakness have "real changes in muscle tone, they have forgotten how to move properly"[71]. Writing in "A Leg to Stand on" The neurologist Oliver Sacks recognized the changes in muscle tone and the organic trigger behind his own experience of functional symptoms. Other neurologist's disagree, according to the Edinburgh based neurologist Jon Stone, "Patients with functional weakness have normal muscle tone and reflexes."[72] Whilst the conscious mind cannot cause one hair of the person's head to turn grey, nor cause an arm to go limp and lose its postural tone it is a mysterious mechanism that allows the hypothetical concept of an unconscious mind to cause a person to be blind, for their legs to go limp or highly selective symptoms to occur anywhere in the body.

Yet traditional models of hysteria require precisely this belief, a belief that is little more credible than saints literally ascending into heaven or limbs growing back if we wish hard enough. In short many neurologists really do believe that medically unexplained symptoms are not organic.[73]

These changes of tone and patterns of weakness do not always follow anatomical patterns. This is key to the diagnosis of hysteria, this "change of tone" and the outwardly observable organic signs should point towards a pathology not directly related to the straightforward wiring of the brain, problems with the brain's memory of movement or the link between haemodynamics and vasomotor activity within the brain for example.

We all learn to move as children; movement is not instinctive. Indeed functional loss of movement is common following injury.

Sacks describes the case of a dog that following injury failed to use its leg even after the fracture healed[74]. With no signals coming from the brain the leg went limp. The owner hurled the poor animal into the sea at which point the swim reflex took over and use of the leg was restored. Instinctive movement is often preserved in hysterical patients and perhaps a key to treatment.

Neurologists refer to this as "distractability", the idea that when the doorbell rings the patient with functional weakness leaps for the door, for example. This is not peculiar to hysteria but a finding in other neurological illnesses such as Encephalitis Lethargica. Indeed Von Economo commenting on the profound new insights that encephalitis had brought states:

> *Hardly ever has the discovery of a disease not only taught us so many separate new facts, but altered our outlook so radically; ... future scientific generations will hardly be able to appreciate our pre-encephalitic neurological and psychiatric conceptions, particularly with regard to so-called functional disturbance.... Now we can describe encephalitis lethargica as a functional affection, but on an organic basis. ... Our disease proves the essential role which quite a number of anatomical structures play with regard to our psychological processes and their arrangements.*[75]

As a disease with a wide-ranging neurochemical and neurophysiological impact, encephalitis lethargica can be compared with few diseases, perhaps to an extent with lyme disease and syphilis. Its role in disturbing the functions of consciousness, our ability to know that we can move for example, obviously is organic. The comparison with hysteria is therefore obvious, symptoms that appear to defy rational medical thought can occur in the absence of localised lesions and they can be explained by disease. As Economo makes clear, functional symptoms, the hysterical signs, in encephalitis lethargica at least, do have an organic basis.

Parkinson's also has its resting and intentional tremors. The French Philosopher Simone Weil found her migraine eased by the distraction of the plain chant at Solesmes. Distractability simply demonstrates the preservation of primitive motor instincts, circuits and organic processes that allow a person to consciously move, to focus and control attention may not be so intact.

THERE IS AN OVER RIDING ASSUMPTION THAT BECAUSE THERE IS NO OBVIOUS PATHOLOGY THAT THERE CANNOT BE ANY PATHOLOGY. Yet, as both Sacks and Hurst make clear, injury to one part of the body can resonate throughout and have a profound impact on the function of the central nervous system. However, even as functional MRI (fMRI) scanners are beginning to show the underlying biological nature of these symptoms, old mechanisms are still implied.

Dr. Patrik Vuilleumier, a neurologist and his colleagues embarked on a study of a number of patients supposedly suffering from hysterical symptoms. In correspondence to the author Vuilleumier states that his work fully vindicates Freudian notions of the unconscious and the nature of hysteria, this raises an obvious question. Did Vuilleumier establish empirically a link between the emotional stressor thought to be the cause of his patient's symptoms and the images he found through the use of SPECT?[h] As mentioned earlier, Halligan has also conducted similar studies that he believes support the diagnosis of hysteria, this time using fMRI. Whether SPECT or fMRI can actually show "emotional disturbances" we need to understand something about the way in which the technology works. We will use fMRI as an example.

fMRI is a monitor of changes in blood flow and blood oxygenation in the brain. When nerve cells are active they consume oxygen carried by hemoglobin in red blood cells through local capillaries. As the brain is activated it requires more oxygen and there is an increase in the blood flow to that area of the brain. This increase is not however instantaneous, there can be a delay for up to 5 seconds. This response typically reaches a peak over 4-5 seconds before falling back to normality. The complexity of this mechanism is demonstrated by the fact that this haemodynamic response will often miss its goal and fall back to less than normal before rising back up again. Therefore the first criticism of any study using fMRI is that you are not actually measuring the function of the brain directly, but rather viewing the mechanism of the brain through another mechanism. It is a bit like star-gazing in a pond.

If the waters are clear and unpolluted then the task is straightforward, if the waters themselves are disturbed then what is actually seen may be very different from what is actually there. Some studies have

---

[h] SPECT: Single photon emission computed tomography

suggested that the increase in cerebral blood flow (CBF) following neural activity is not even directly related to the metabolic demands of the brain region, but rather is driven by the presence of neurotransmitters, especially glutamate. So that if the waters were not murky enough, claiming that any fMRI study can vindicate the work of Freud and a purely psychoanalytical view of the brain is unlikely.

In hysteria, all fMRI or SPECT can ever reveal is the organic basis of physical symptoms. It introduces a host of possible organic causes, possible problems with neurotransmitters, a fault with the mechanism that controls the increase and decrease in cerebral blood flow. It certainly does not show an unconscious mind preventing an individual from moving their arm whatever Vuilleumier might say. Rita Carter writes in her book *Mapping the Mind,* "The vision of the brain we have now is probably no more complete or accurate than a sixteenth-century map of the world."

These neuro-imaging studies are certainly beginning to uncover the organic basis behind functional disorders of the brain but their interpretation is a different matter. Vuilleumier's study dealt with hysterical paralysis and showed "functional" abnormalities[i] to striatothalamocortical circuits. What is striking is that fMRI and SPECT studies show organic signs as a predictor of recovery. They show recovery dictated by the level of blood flow in the contralateral caudate and thalamus. Significantly decreased blood flow in these areas predicts poor recovery.

---

[i] "Functional Abnormalities", in other words these parts of the brain were not working though the organic reason could not be determined by the technology being used.

In contrast Helen Crimlisk and her colleagues point out that receipt of state benefits rather than blood flow is a good predictor of poor recovery,[76] the patient being unwilling rather than unable to improve for fear of financial loss. This is an echo of the notion of secondary gain and the DSM idea that patients actually benefit from being ill. This of course does not recognise the degree of disability and level of evidence needed to actually obtain these benefits in the first place. Nor does it recognise that these patients may well have greater physical disability in the form of decreased cerebral perfusion, an organic predictor, than those who do not receive state aid.[j] That many patients with unexplained somatic symptoms suffer secondary loss to a greater extent than they experience secondary gain is discussed in the case of ME and "CFS" by the Canadian psychologist Donald Dutton in his work "Depression/ Somatisation Explanations for the Chronic Fatigue Syndrome: A Critical Review."[77]

What neuro-imaging does not show is that an emotional event has actually caused these fMRI/SPECT anomalies. It does not reveal that in a stressful moment a patient thought they might be having a stroke or going blind and that this idea of illness caused them to have stroke like symptoms or lose their sight. It may well be the case that without the label of disease the increasingly desperate patient's presentation of illness assumes those dimensions that conform to their own understanding of anatomy and cultural pressures. That in effect, their presentation becomes increasingly divorced from

---

[j] One direct result of Sharpe & Wessely's erroneous portrayal of Myalgic Encephalomyelitis as an illness in which patients are "fatigued" and their use of the odd term "Chronic Fatigue Syndrome" has been that even bed-bound patients often find it impossible to claim any kind of state benefit- Hooper et al discuss this extensively- "The mental health movement, persecution of patients?"

scientific notions of pathology.[78] But this does not mean that there is no disease.

The striatal areas of the brain shown in Vuilleumier's study are also involved in motor neglect as lesions on conventional MRI show.

Motor neglect is a common term in neurology; it means that the wiring that allows a patient to move an arm or a leg is intact but that there is a lesion, damage visible on MRI or CT scan to the part of the brain that enables the person to know that they can move. The over-riding message is, as Carter suggests, that the brain is a complex organism. To use blood flow in the brain, a mechanism that is in itself prone to fluctuations, as a tool to suggest that an idea has caused illness is a difficult leap of faith. Using it to diagnose what Stone and colleagues define as "neurosis"[4], functional weakness or other functional symptoms and to call these non-organic is a leap in the dark. Streeton highlights the complexity of cerebral perfusion and its role in possible missed diagnoses in his book *Orthostatic Disorders of the Circulation,* Streeton states:

> *The major complaints of about one-third of patients were fatigue and weakness, often overshadowing lightheadedness, which was not even mentioned until its presence was specifically enquired into. Because fatigue and "weakness" are such common, relatively nonspecific symptoms that do not lead physicians to suspect the presence of orthostatic hypotension, few of these patients had their blood pressures measured while they were standing. Many had gone through extensive laboratory tests with entirely negative results, leading to an erroneous diagnosis of psychoneurosis.* [79]

Yet rather than reflect on possible organic causes other than lesions in the striatal areas of the brain both Vuilleumier, and Halligan in his own study, suggest that these fMRI or SPECT abnormalities are evidence of psychological deficits, that they support the notion that emotional rather than organic changes are not only the cause but the perpetuating factor. Whilst clearly demonstrating that the symptoms suffered by their patients are not caused by gross damage to the central nervous system they fail to consider the constellation of possible organic causes that might lead to a patient not knowing that they can move.

We are therefore reminded again that functional symptoms are diseases affecting the function of consciousness as surely as syphilis or Lyme disease do. As we have already noted, the French neurologist Charcot suggested that "dynamic lesions" were the cause of functional symptoms such as weakness and paralysis. At first he felt that these must be caused by disease but without even x-ray or MRI at his disposal, he gave up.

Influenced by Thomas Briquet and Sigmund Freud he concluded that these symptoms and signs were not organic in origin. Charcot pioneered many areas of medicine but the techniques he used to diagnose hysteria often resulted in serious missed diagnoses. Effectively he continued medicine's failure to recognise its own ignorance. Even though hysterical patients may present to a neurologist in a way as distinctive as Parkinson's and migraine, with functional symptoms there is considerable blurring between explained and unexplained disease. Stone and colleagues suggest that 64% of patients with functional symptoms will have quantifiable organic illness. Yet the actual signs of hysteria have a long tradition and one that despite research is not easily altered:

> *In an attempt to validate recent assertions that the strongest indicators of hysteria are the "positive" findings in the neurological examination, seven of the most accepted features (history of hypochondriasis, secondary gain, la belle indifference, nonanatomical sensory loss, split of midline by pain or vibratory stimulation, changing boundaries of hypalgesia, giveaway weakness) were sought in 30 consecutive neurology service admissions with acute structural nervous system damage. All subjects showed at least one of these findings; most presented three or four. The presence of these "positive" findings of hysteria in patients with acute structural brain disease invalidates their use as pathognomonic evidence of hysteria. A second, retrospective study on the misdiagnosis of hysteria demonstrated that women, homosexual men, the psychiatrically ill, and patients presenting plausible psychogenic explanations for their illness are most liable to be misdiagnosed. Certain disorders, particularly movement disorders and paralysis, are most often mislabeled as hysteria. A diagnosis of hysteria must be made with great caution as it so often proves incorrect:*
>
> *The validity of hysterical signs and symptoms.*[80]

Here Gould and colleagues clearly warn that these positive signs are no sure way to diagnosis. Yet even though Gould finds these signs in patients with acute structural brain disease, many neurologists continue to argue that these signs are still not organic. For example some may argue that a patient with acute structural brain disease presents these signs as a reaction to the disease process, abnormal illness behaviour or elaboration, they are not actually caused by the disease process itself.

What Gould also highlights is of course the social nature of these symptoms. It is not "Joe Average" who is most likely to be misdiagnosed but intellectuals, homosexuals and those already deemed to be insane. Irrespective of neurology's ability to explain away the signs, the stigmata of hysteria, the fact remains that they are a stigma most often found in those on the fringes of society or indeed those obligingly able to save time and a few pence on tests through a rational explanation of their own condition. Gould reminds us again that when tests are negative the deviance of the patient from social norms does lead to an increased risk of misdiagnosis.

Neurology is therefore by its very nature, its obsession with structural disease, geared towards spectacular missed diagnoses. It pursues a spurious taxonomy of illness in which Tourette's is neurological whilst schizophrenia is not. In an age when fMRI has begun to show the organic basis of these symptoms, many neurologists continue to assert that functional weakness is the same as psychogenic weakness. Even worse, neurology has failed to learn the lessons of history, that missed diagnoses can and do occur. Heaving the diagnosis of hysteria out of its Freudian and psychiatric routes and accepting that what are currently classified as "functional symptoms" have a myriad of complex organic causes will never happen whilst psychiatrists are left to manage this group of patients. Psychiatrists are trained to look for emotional and social causes and thus for them, functional symptoms are ultimately psychogenic.

psychogenic- originating in the mind or in mental or emotional conflict- OED(2007)

4) Studies of hysteria

> *Reinvestigation of these patients is both expensive and potentially dangerous and should be avoided where no clear clinical indication exists.*[81] Helen Crimlisk (1998)

> *At follow up during 1996, subjects underwent a semistructured interview designed to assess the evolution of the index symptom, the occurrence of other somatic or psychological symptoms, the subjects' utilisation of medical and psychiatric services, and details of any state financial benefits received. The schedule for affective disorders and schizophrenia was completed for each subject, and this was supplemented by all available hospital and general practitioner records. Current and lifetime diagnoses according to the 10th revision of the international classification of diseases were obtained. Subjects were reassessed physically by a neurologist.*[82]
>
> Helen Crimlisk (1998)

In law an individual is assumed innocent until proven guilty. In medicine there is an underlying assumption that when tests are negative a patient must be somatising until proven innocent, usually via further tests. More bizarrely, in medicine a patient with unexplained disease is considered to be somatising, even when disease is discovered if, in the clinician's opinion, this disease does not fully explain their presentation. Crimlisk includes within her study of patients with functional symptoms those who do have neurological conditions but which are considered not severe enough to explain all their symptoms. She fundamentally assumes a weakness on the part of the individual. There is an assumption that frequent attenders must be worried about their health. There is no consideration that this group of patients might be forced to attend by the inadequacy of care offered for their health.

Crimlisk assumes that amongst the records of patients with functional symptoms there must be somatising, psychological symptoms, a long history preceding the grand episode of hysteria.

Worst of all is the assumption that a neurologist by applying the same techniques first devised by Charcot and Babinski can infallibly diagnose an individual. She does not entertain for a moment that there is a hidden history of missed diagnoses, botched interventions, inaccurate records, even incorrectly entered ICD codes (international classification of diseases). Most importantly of all she does not recognise that in some cases at least missed pathology might disturb the brain's cognition of a limb and thus a person's ability to move.

As the neuropscyhologist Alexander Romanovich Luria recognised, the body is one complete organ[83]. A pebble of illness thrown in a pond can have ever spreading ripples. In Crimlisk's study nearly half of the patients had a history of organic neurological illness.

Faced with medically unexplained symptoms in military casualties, the American Civil War doctor Weir Mitchell writes:

> *We have some doubt as to whether this form of pain ever originates at the moment of the wounding. . . Of the*

special cause which provokes it, we know nothing, except that it has sometimes followed the transfer of pathological changes from a wounded nerve to unwounded nerves, and has then been felt in their distribution, so that we do not need a direct wound to bring it about.[84]

In diagnosis symptoms that do not fit are often left unexplained in the shadows. Even worse the patient is often told "it is all functional" (as they collapse from cardiac related syncope as did the author for example).

Focused investigation of these anomalies is simply not routine. Faced with bewildering and inconsistent symptoms the cultural and economic force is not to pursue further tests but to leave the patient untreated and distressed with a diagnosis that few understand or are able to challenge. They are left with simplistic cartesian explanations such as, "this is a software problem."

Every age uses images to understand the human body. In previous centuries the body was a machine. In our age the body is seen as a computer with software and hardware. Functional symptoms are seen as software problems. What in Freud's time was seen as a psychosis disturbing the body is re-interpreted and sanitised for the modern era. It is made "acceptable" for the patient.

This would be bad enough but even in the very breath with which Crimlisk states that considering a patient as a somatiser does lead to missed diagnoses she also states that re-investigation of these patients is not acceptable. In a health service with limited resources it is interesting that Crimlisk plays the financial rather than the political card. Rather than question why resources for investigating these patients are limited she assumes that such investigations are worthless and detrimental in the first place. There are strong Freudian overtones to this, a fear that medical intervention will re-inforce "illness beliefs", that it will cause the individual to even further repress the psychological dimension of their condition. Crimlisk clearly wishes for the patient to admit "thank you doctor, I now see it all, my father's death and my mother's o'erhasty marriage, this was the cause of my lunacy." Crimlisk would do well to reflect on the words of Slater:

> *The diagnosis of 'hysteria' is all too often a way of avoiding a confrontation with our own ignorance. This is especially dangerous when there is an underlying organic pathology, not yet recognised. In this penumbra we find patients who know themselves to be ill but, coming up against the blank faces of doctors who refuse to believe in the reality of their illness, proceed by way of emotional lability, overstatement and demands for attention ... Here is an area where catastrophic errors can be made. In fact it is often possible to recognise the presence though not the nature of the unrecognisable, to know that a man must be ill or in pain when all the tests are negative. But it is only possible to those who come to their task in a spirit of humility. In the main the diagnosis of 'hysteria' applies to a disorder of the doctor–patient relationship. It is evidence of non-communication, of a mutual misunderstanding ... We are, often, unwilling to tell the*

> *full truth or to admit to ignorance ... Evasions, even untruths, on the doctor's side are among the most powerful and frequently used methods he has for bringing about an efflorescence of 'hysteria'.*[85]

Slater recognizes that hysteria, when seen as a psychiatric condition, results precisely from the ways in which medicine seeks to avoid admitting to the patient that it does not have the answers. Furthermore, he recognizes that it is precisely by refusing to publicly admit ignorance that the patient will further increase their efforts to convince the doctor that they are ill. Contemporary neurologists call this "illness behaviour." The irony is, as Slater recognized that the illness stems from the physician, not from the patient.

Indeed the psychologising of basic human experiences and their random association with illness is all too often used to explain disease. As Slater realised, such occurence is often mere chance. Alongside stressful life events one symptoms stands out. In neurology and psychiatry "dissociation" is often claimed to go hand in hand with functional symptoms.

Dissociation is when a person feels disconnected from themselves, their body or their environment. It is therefore loaded with dualism as the person must actively participate in the feat of dissociation from apparently healthy neural mechanisms within their brain, mechanisms that allow them to feel a limb as part of themselves and then sustain this dissociative action without conscious awareness that they are doing so. It is seen as an important part of the diagnosis of for example, functional weakness, and a sign that the patient has indeed experienced a stressful event, a panic attack or other "episode" even though the patient themselves might be completely unaware of this. The interpretation of what the experience might actually mean, what emotions might be involved, is not left to the patient, rather, retrospectively the doctor tells them what this experience meant.

The problem is as Stone acknowledges in his paper "Dissociation: what is it and why is it important?"[86] that this symptom can occur quite frequently in disease, migraine, and epilepsy. It is not pathognomonic [k] of a psychogenic disorder yet he is still prepared, when tests are negative, to use it as a diagnostic tool. Indeed "dissociation" can occur in any condition where there has been disuse of the haemodynamic processes in the brain that are required for neural activation.

It occurs following immobilization of a limb due to fracture, though such cases have declined since 1974 when long-term immobilization of fractures ceased. So that in effect what is experienced is a disturbance of the brain's usual image of itself. To describe the experience as a "panic attack" may well, in patients who do experience panic, be a reaction to the experience, not the initial cause, a point Stone himself acknowledges. Stone describes 2 cases; one where epilepsy is suspected but the EEG proves normal though QEEG[l], which has a higher beam resolution, is not offered nor

---

[k] Pathognomic- an indication of a specific disease
[l] QEEG has been used in the USA in legal cases. The reader is directed to the QEEG report by Prof. Frank Duffy in the notes as detailed earlier and to consider the issue for themselves.

seemingly is any cardio-vascular workout undertaken. The recurrence during a stressful hospital appointment could point to a vascular anomaly rather than plain stress for example. The second is that of a student, apparently to all intents and purposes intensely daydreaming. At any rate the notion that such moments are in themselves indications of neurotic episodes[87] receives little respect from the poet Yeats:

> *Our Master Caesar is in the tent*
> *Where the maps are spread,*
> *His eyes fixed upon nothing,*
> *A hand under his head.*
> *(Like a long-legged fly upon the stream*
> *His mind moves upon silence.)*
>   -Long-Legged Fly- William Butler Yeats

Yeats describes in the same poem the intense creativity of Michael Angelo "With no more sound than the mice make, His hand moves to and fro," so that "blank spells", which in the case of the student, Stone associates with "panic" are, in Michael Angelo's case at least, moments of intense connection. That one needs to be actively thinking to be concentrating, as Yeats recognizes, is nonsense. No doubt Caesar arose from his map feeling "strange" but that would merely indicate a move back to the world around him, a change in the focus of his mind and resolution to act rather than anything like depersonalization or derealisation. If anything, both Caesar and Michael Angelo are profoundly connected to the world in a way that may well have come as a shock to the student described by Stone.

Describing "blank spells" as dissociative states when, as Yeats recognizes, they can be moments of intense brilliance and creativity demonstrates again the disastrous collision between neurology as a Science and the actual reality of human experience. As a history of science shows, when science has moved from descriptions of reality to interpretations of what they might mean, it is the interpretation that so often proves incorrect.

Of course, as Stone suggests, there is indeed a biology of "blank spells" but if that is so then suggestions that a patient subsequently developed weakness because of an idea that they had developed disease is to call yet again upon mysterious mechanisms rather than an organic process. It merely postpones rather than solves the problem.

It may well be that symptoms of weakness, that can currently not be explained by medicine, do indeed stem from plasticity within the brain, an ability to "rewire" itself or use redundant connections in odd ways, what Sacks describes as a "third realm" of neurology. Nature abhors a vacuum so that if a person experiences a sense of disconnection from a part of their body then organic explanations must be rooted in the tendency of the cortex to rewire itself as a result. Patients with both organic and unexplained symptoms often describe this as a feeling that someone has literally ripped out all the wiring and randomly reconnected the circuits. That there is such a "third realm" was Sacks' conclusion following his own experience of disconnection[m] from his leg after breaking it. It is also an idea

---

[m] I have used the word disconnection deliberately as it reflects an organic rather than a psychological process. The manner by which a person may feel

supported by the work of Ramachandran and crucially does not rely on ideas of illness.

This would be all well and good but as Crimlisk's study shows, whatever the reassurances, there is an underlying assumption that this *third realm* is one well and truly anchored in the psychiatric and the disordered self. Sacks himself is at pains to point out that his experience differs from that of the "hysterics." One wonders what history of somatising lurks in Sacks' medical files ready for Crimlisk or others to discover. According to Crimlisk's approach any unexplained somatic symptom would have been of note. Despite claims for neutrality of the term "functional" and Sacks' painstaking account of the organic basis of his experience, contemporary neurology is not always so astute. Stone and colleagues suggest functional symptoms occur following accidents in "vulnerable individuals."

It is not the case that neurologists set out to deceive; they really do believe that functional symptoms are not caused by disease. Yet surely the systematic mapping of a whole patient, their systems, organs, psychology, every aspect of their physiology, surely this should be the object of any doctor faced with medically unexplained symptoms?

---

disconnected from a limb is probably vast. Injury, for example, may produce a heightened awareness of the limb that is deemed "abnormal" to the brain. Subsequently the brain may rewire itself in search of what it deems to be normal, like a weed randomly seeking out sunlight. The inconsistencies found in "hysteria" and indeed in phantom pain syndromes may therefore stem from this "seeking out" of normality, an organic rather than a psychological process that has nothing to do with the "idea" of illness as suggested by both Stone, Charcot and others.

It cannot be ignored that in recent years misdiagnosis has been comprehensively studied. A systematic review of misdiagnosis based on 27 studies and 1500 patients found a misdiagnosis rate of around 5%. In one influential study patients were reassessed via a quiz through the post. One could imagine conducting a similar study in ancient Greece or Rome and finding exactly the same statistic. Why? Because diagnosis is based on a society's definition of disease and its knowledge of the disease process. This rarely changes in the course of a few decades. Diseases once considered hysterical, such as camptocormia[n], take a long time to be accepted as organic. Even worse, once diagnosed as suffering a "functional" disorder, such as functional weakness, access to further tests is often barred and the whole condition given this label. A quiz through the post is hardly a complete technological re-examination of a patient. Asking a busy GP is even less likely to reveal any startling insights not least as Psychiatric diagnoses are very sticky. As Rosenham discovered ("In being sane in insane places") once inside the asylum it is very hard to get out, no matter how sane (or organic) the patient actually is.

David Rosenham was an American psychologist who earned his master's degree at Columbia University in 1953. Rosenham was deeply suspicious of psychiatry and the ways in which it diagnosed patients. As a result he devised an experiment in which he and his colleagues pretended that they kept hearing the word "thud" in an attempt to gain admission to 12 different psychiatric hospitals in 5

---

[n] Originally considered a psychogenic disorder, camptocormia presents as an abnormal posture with marked flexion of thoracolumbar spine that abates in the recumbent position, it is becoming an increasingly recognized feature of parkinsonian and dystonic disorders. In other words it was once thought to be "all in the mind" but is now seen as organic.

different states across America. Apart from false names and personal details this was the only lie they were allowed to tell, in all other respects their behaviour was to be completely normal. All were admitted, diagnosed with schizophrenia, treated and eventually discharged with the diagnosis that their schizophrenia was in remission. During their stay staff interpreted much of their normal behaviour as evidence of mental illness. For example the note-taking of one pseudo-patient was listed as "writing behaviour" and considered pathological. Rosenham also designed an experiment in which hospital staff were told that a number of admissions were pseudo-patients and asked to identify which patients they were. Out of 193 patients, 41 were considered to be imposters and a further 42 were considered suspect. In reality Rosenham had sent no pseudo-patients to this particular hospital.

Rosenham therefore eloquently demonstrated that where value judgements are used to make a diagnosis these can be deeply flawed. With regard to medically unexplained symptoms Rosenham's work suggests that in the absence of positive laboratory tests, who has developed physical symptoms as a result of neurosis and who has not can be a deeply contentious question. Rosenham also demonstrates that once a diagnosis is made it is very difficult to have it overturned. His pseudo-patients were told their Schizophrenia was in remission, not that they were wrongly diagnosed or even cured.

Evidence suggests that once a diagnosis is made subsequent physicians are more likely to stick rigidly to the label irrespective of any symptom that does not fit. Some neurologists have even suggested that further testing is likely to be detrimental to the patient's condition whilst at the same time urging NHS managers that an early "functional" diagnosis is a very effective cost cutting measure. Patients are even discouraged from talking to each other and support groups viewed as encouraging "illness behaviour."[o] Those patients referred for Cognitive Behavioural Therapy are "press-ganged" into reporting improvements or face discharge[88]. In reality it is very difficult for a patient to challenge the hegemony of their doctor. To do so is to be a difficult patient and 66%[89] of these are alleged to have a psychiatric disorder.

In this culture of certainty it is perhaps worth remembering that a New Scientist article cited that 39%[90] of patients in intensive care have serious problems that are missed and only discovered by autopsy. In the case of hysteria, unless the patient inconveniently dies, any errors are unlikely to be uncovered. Disease is not always progressive. Disease does not always affect the ability to move it can also affect our ability to know that we can move.

---

[o] A Neurologist's e-mail CC to myself and Dr. David Bateman

5) Mechanisms of Compliance

> *Central to recent debates about hysteria and conversion disorders is the extent to which a person's illness presentation is considered a product of free will and hence social deviance or the result of psychopathology and/or psychosocial influences beyond the volitional control of the subject-*
>
> —*Dr. Chris Bass*

One Canadian physician, Byron Hyde, famed for his work with patients suffering from myalgic encephalomyelitis states that when faced with a patient for whom he is unable to determine a physical cause he blames his failings as a doctor and the limitations of technology. Bass does the reverse, irrespective of his own personal views, by highlighting the notion of free will he raises an obscene spectre, that the patient is culpable. They are in some way responsible for their illness through either lacking the volition, the will to move, or by unconsciously exerting a will not to move. Bass's colleague Vuilleumier echoes this in his interpretation of the SPECT studies of hysterical patients. It is no good to assert that the patient is suffering from "psychosocial influences" beyond their control. Bass could say the same of cancer patients, heart disease or all manner of illness.

To state that hysteria is a psychosocial illness is meaningless (there is a psychological and social dimension to all illness) but that does not constitute the illness itself. To Bass the patient is socially deviant because their presentation does not fit with the neat social classes of disease that medicine has prepared for them. As far as medicine is concerned the so called "hysterical" patient is like a mongrel dog at the line up for Crufts. They simply do not fit with classical medicine and for that reason alone they are excluded.

The strategy then becomes to persuade the patient that they do not fit into the murky and shifting categories of organic illness, in essence to "sell" them a diagnosis. One leaflet entitled "what is functional weakness" states "unless you accept your diagnosis you are unlikely to get better." Irrespective of the obvious "faith healing" echoes of this statement and its biblical requirement to believe in the healer, there is yet again an element of control, in medical-speak "a management issue". Admittedly patients who become stressed by the diagnostic process are likely to worsen their condition, but this is true of any condition. If symptoms are "real" then the emphasis should be on the "cure" not acceptance of the diagnosis itself. If explanations for functional symptoms echoed the reality of patient experience, there would be no need for neurologist and psychiatrist to embark on a mission of religious conversion to frighten the patient into believing and accepting what they have been told.

Saying, "unless you accept your diagnosis" is rather like declaring, "unless you accept you have the flu you are unlikely to get better." Rather than remove the individual from the cause of their symptoms it implicates the patient, yet again, as having a responsibility to do as they are told, fall into line. The ultimate treatment for deviance is confinement to a psychiatric ward as described by Fiona Whelpton in "the cycle path" and as described by Karen Armstrong in "the spiral staircase." There are other more subtle methods of control with the physician as "gate-keeper" to further tests and "preventer" of inappropriate consultations. This attempt at control is not limited to simply manipulating patients. The Newcastle neurologist Dr. Philip Nichols stated in one letter to a colleague that surgery for fecal incontinence would be mutilation. Another letter clearly castigated the judgement of the head of clinical psychology (and head of the local CFS service at the time) that the patient (myself, the author of this book) was suffering a post-viral illness and insisted on a "proper" psychiatric assessment.

Nichols was warned that his course of action could constitute a breach of human rights by the patient's own GP Dr. Sarah Goodman, "whatever the cause of the incontinence I feel he has a right to explore possible management strategies." To which Nichols replied, "I do not think that this is a logical approach and I feel will only exacerbate the psychological problems." Had Nichols pursued bowel and bladder investigations he would have found a grossly abnormal neurogenic bladder with marked detrusor overactivity and a loss of sensation in the bowel consistent with local autonomic dysfunction.

Whilst Nichols insisted that the patient's condition was wholly psychological and that it had no organic basis whatsoever the patient went on to have multi-site photoplethysmography tests, a measure of peripheral blood flow and autonomic function. These proved abnormal. He also had 24hr blood pressure and heart rate monitoring which proved, abnormal. Why should it be that his "weak side" should be so freezing cold when camped at altitude in the Tien Shan if not caused by an autonomic vascular defect? The patient had "systemic autonomic dysfunction." Indeed one expert described the patient as having, "obvious signs of autonomic neuropathy" that Nichols had missed.

Complex diseases remain "out of reach of the scanner" precisely because "the scanner" is actually ineffectual in detecting the pathology of most disease.

That the approach of physicians such as Nichols is questionable is hinted at by the fact that according to the Nightingale Research Foundation, which has over 23 years experience in diagnosing patients with myalgic encephalomyelitis, 90% of patients with ME/CFS have missed diagnoses only revealed by total body mapping rather than the "hunch" driven approach of traditional medicine. That in neurology at least this "hunch" driven approach frequently settles on using the so called "signs" of hysteria as a diagnostic tool, as an infallible means of deducing that there is no organic basis for a patients symptoms, is likely to remain a questionable mix of "professional guesswork" and the religiosity and adherence to creeds that all organizations exude, not least the British Association of Neurologists. This creed is at its most deadly in the complex interplay of disease that causes physical and neuropsychiatric symptoms, for example, the cases of Ms. A and Mr. C.

Ms. A's case was published in the American Journal of Psychiatry in 2002 and detailed the sad story of how she had presented with a range of puzzling symptoms that under hypnosis apparently seemed to improve. This appeared to confirm a psychiatric diagnosis as indeed did other circumstantial events. When her condition steadily worsened help and treatment were declined, the creed being that patients with conversion disorder will have their illness beliefs increased and thus their condition worsened by any organically motivated intervention. When Ms. A was unable to feed herself the American psychiatrists treating her relented in their non-organic approach and began feeding Ms. A through a nasal gastric tube. She repeatedly removed the tube and when she did eat a meal normally, the nursing staff assumed that her previous reticence had been due to neurosis. This proved disastrous for Ms. A, she proceeded to choke whilst eating but on examination no food was seen in her throat, further confirmation that her condition was functional and therefore hysterical. She was found later sat in her chair dead; she had suffered a cardiopulmonary arrest. She was resuscitated but in intensive care they found her to be brain dead. The arrest being caused by choking associated with brainstem disease.

It was at this point that the psychiatrists noted the results of an EEG that had been taken in the days before her death. The EEG was abnormal and suggested "a rapid neurodegenerative process." A full autopsy was instigated during which it was found that Mrs. A had died from choking on her food. Though visual examination of her brain revealed no abnormalities whatsoever, under the microscope there was extensive pathology caused by prion infection. The terrifying conclusion was made that Ms. A was suffering from Creutzfeldt-Jakob disease. Not only had this been missed during her lifetime, but treatment that could have eased her suffering had been withheld for fear that she might just be convinced that there was something physically wrong with her.

Faced with a case that could possibly undermine the whole edifice of "functional illness" Ms. A's psychiatrist wrote:

> *Given the changes seen in Ms. A's personality, it would be expected that she would have had a diminished capacity to use her usual coping strategies at a time of extreme duress. Because of this, we believe she communicated distress through an elaboration of symptoms (illness behaviour) during the course of her illness.*[91]

Whether they felt that death was the final elaboration of her symptoms is not made clear. It was not long before they found support from Edinburgh neurologist Jon Stone, psychiatrist Michael Sharpe and Martin Ziedler an expert in Creutzfeldt-Jakob disease. This took the form of a letter to the editor of the American Journal of psychiatry and was written even though neither Ziedler or his colleagues had ever met or examined Ms.A whilst still alive:

> *It would be a shame, however, if this case conference reinforced the erroneous idea that the development of neurological disease in such cases is the norm. Failure to make a positive diagnosis of conversion disorder can have serious adverse consequences. The patient may be denied appropriate treatment and management that vitally depends on persuading him or her that the symptoms are reversible and not due to disease. We should not withhold the diagnosis simply because we occasionally get it wrong.*[92]

It would be a shame indeed, for not only had a similar misdiagnosis been reported in the journal "Psychosomatics" in June 1999[93] but the case of Mr. C described in that journal clearly highlighted the complex behavioural changes that take place in prion disease. Mr. C presented with a puzzling mix of behavioural changes and found himself treated with psychotropic drugs, counseling and ultimately electric shock therapy. Quoting a variety of established sources including a "Textbook of Neurology" the authors of that paper state:

> *In fact, there is a well-described prodromal phase of the disease in which symptoms may be vague and can include weight loss, fatigue, dizziness, headache, disorders of sleep, impaired judgment, and unusual behavior. An unusually intense emotional response to the environment as well as delusions, hallucinations, and agitation may be interpreted as a depressive or psychotic illness.*

So that one would expect a lesson to be learned, that the complex detective work of hunting for an organic cause in anyone presenting with sudden personality changes or medically unexplained symptoms should be pursued rigorously and without recourse to theories of somatisation.

There are obvious conceptual problems concerning the nature of disease in much of Stone and colleagues' letter to the editor. For one thing, Ms. A's brain was damaged by CJD and thus she was compromised in her ability to function due to organic, not emotional causes. As the case of Mr. C makes clear this is indeed a factor in

CJD, albeit a poorly documented one. References to her sexuality and same sex relationships in the same context as mentioning psychosocial stressors reveal more about the prejudices of her psychiatrists than the nature of the disease she was suffering. That the patient may be denied "appropriate treatment" and treated with "aggressive behavioral strategies to treat a presumptive conversion disorder" goes to the heart of the problem, a violence against the person when organic explanations fail to emerge.

Not only do Stone and colleagues apparently fail to recognize that Ms. A's case was a misdiagnosis of monumental proportions but it is unclear whether they recognize the iatrogenic damage she suffered at the hands of her physicians. By default rather than conscious will, they encourage others to mete out the same treatment to any patient who has symptoms that cannot be explained by the physician's knowledge of disease. In essence this is to withhold any treatment, assistance or pharmacological intervention that has an organic goal and to pursue ruthlessly the sympathetic magic of psychiatry and conversion disorder. In Ms. A's case this was this was a management strategy followed even up to the final moments of her life.

The patient must heed the authority of their physician no matter how frankly insane and damaging the wielding of that authority might be. When the patient is too ill or lacks the education to challenge the rationale of such diagnoses then their use is tantamount to intellectual abuse and exploitation. That such a path should be followed without question by the likes of Stone, Sharpe and colleagues has more to do with the received wisdom that even when every cell in the patients' body has been placed beneath the microscope, there will remain conditions unexplained by disease.

That Ms. A's condition could only be explained under the microscope and that no doubt many other conditions await even more detailed study will forever remain the folly behind a diagnosis of conversion disorder.[p]

---

[p] The last part of this chapter is based on research that the author originally forwarded to Richard Webster, author of "Why Freud Was Wrong." Richard drafted an essay including the case of Ms.A on which the final part of this chapter is based. At time of writing Richard's article is available on www.RichardWebster.net. I have also used a number of other sources from this excellent, though unpublished, article.

6) Hysteria as failed communication

In a survey of broadsheet newspapers Stone, Colyer, Sharpe and colleagues analysed contemporary usage of the word psychosomatic,[94] they write:

> *The word "psychosomatic" is used frequently in newspapers to mean an illness that is not important or is imaginary, malingering, a sign of madness, or a character flaw.... Psychiatric stigma may take on particular importance for patients with physical symptoms. Patients with physical symptoms that are unexplained by disease may feel rejected if they are told that their problems are psychosomatic... Some have even wondered whether it is ethical to transmit a diagnosis such as "psychosomatic illness," which can have such damaging implications for the doctor-patient relationship. We also found that when its use was not pejorative, the word "psychosomatic" was usually used to suggest a psychological problem or the effects of the mind on the body rather than a reciprocal interaction between body and mind. Some patients may find this meaning also stigmatizing.*[4]

It is in this article that perhaps a fundamental failure of communication has taken place. Stone and colleagues fail to recognize that words have pragmatic not just semantic value. Thus there are good reasons why the media sees diseases of the mind as less important or imaginary, it is because in medical terms they are precisely that, phantoms. Indeed if we are to consider "a reciprocal interaction between body and mind" then we must ask what precisely is meant by "mind"? The term psychosomatic is thus essentially Cartesian dualism. Descartes assumed that the pineal gland was the part of the brain that connected the brain to the soul. Modern neurology, with its focus on lesion localization, has done away with such myths and replaced it instead with notions of disordered executive function. It is therefore executive function that becomes the seat of the soul. The essence of who we are is seen in our moral rather than physical ability to act.

The Californian neuropsychiatrist, Dr. Jay Goldstein, calls bodily complaints that have no structural explanation "neurosomatic." All symptoms must have their flowering in the brain, it is the mechanism by which we know we are ill. Goldstein argues that where, for example, pain is felt out of all proportion to any rheumatic explanation as in fibromyalgia, it is brain dysfunction, a disorder of chemicals and complex processes that is at fault. It is not some phantasm called "the mind". We know we are ill because there is a disorder in the mechanism by which we know we are ill. The aim of treatment must be to biologically assess and treat this imbalance either through the adminstration of neurotransmitters or, one treatment favoured by Goldstein, infusions of lidocaine. To look at the brain as an organ remarkably different from the heart or liver is to privilege sentience and consciousness as somehow different from the organ they arise from.

As the Hungarian psychiatrist Thomas Szasz recognized, there is no such thing as mental illness, there are just people and life situations.[95] Concepts of individual responsibility in illness are based on "fake disease" concepts. Clinical depression is a problem with serotonin, Obsessive Compulsive Disorder is closely related to the neurological mechanisms of Tourette's. In other words they are organic. So that to rehabilitate language that has undergone such a culture shift is a naïve pursuit, no matter how noble the intentions. When Yeats ended his poem Lapis Lazuli with: *"Their eyes, their ancient, glittering eyes, are gay."* He had little idea that the word would undergo an irreversible shift in meaning over the next century. It continues to do so with its current "youth meaning" often being something that is useless, pathetic rather than any sexual connotation. Such a shift changes the reading of Yeats' poem out of all recognition. The same is true for hysteria, somatoform, psychosomatic, and even the word functional, the latter is routinely abused within the medical profession itself. As noted in previous chapters it can mean a disease affecting the function of an organ or to indicate a mental illness. Often users fail to explain exactly which semantic meaning they intend. To attempt to communicate with patients through language that has seen such a cultural shift is likely to exacerbate problems between doctor and patient. As noted earlier the word "functional" has no universal meaning within the medical profession and is often seen as simply mental illness.

The entire notion of Wesselyian psychiatry of which Sharpe and Carson are frequently seen as members, is the rehabilitation of pre-Freudian notions of mind and body, the reciprocal interaction between body and mind. If these physicians aimed to treat both mind and body as equal partners then the notion would certainly have to be applauded. As Wessely has stated "all illnesses have a psychological factor. It would be an odd world if they didn't"[96]. Of course that does not mean that psychogenic factors or neurosis are good or accurate descriptions of the illness itself. Faced with the opportunity to research the organicity of these symptoms, research has focused on symptoms research and psychiatric causation, it has neglected physical signs and the high level of co-morbidity. That concerted biomedical research can uncover the organic basis of symptoms in complex illness is shown by the work of Hyde, Cheney, Goldstein and others[97] in their work on both Chronic Fatigue Syndrome and Myalgic Encephalomyelitis. As Hyde himself has stated, "For me, a patient with an initial diagnosis of ME and CFS can be a gold mine of disease, missed injuries, physical and physiological anomalies, and genetic curiosities"[98]. This contrasts with Wessely's comments that ME and CFS "have no anatomical or physiological basis"[99].

If 19th and 20th Century concepts of medicine were based on identifying individual pathologies, after the manner of Osler father of modern medicine, then the 21st requires a more robust approach to understand the complex processes underlying functional illness. By comprehensively mapping an individual's systems and organs Hyde has found missed diagnoses and organic anomalies in 90% of patients with CFS and ME. Many of these patients had previously been told there was nothing physically wrong with them.[100]

By explaining physical symptoms through language that no longer satisfies the reality the patient experiences, the scene is therefore set for precisely the break down of doctor and patient relationships[101] that the writers fear. In patients who experience symptoms that disturb mental functioning it is an obvious statement that they are more likely to accept a mental explanation with a comparable decline in acceptability as their experience of symptoms becomes more organic.[102] What is clearly needed is a paradigm shift to engage with the organic manifestations of medically unexplained symptoms. If this is not taken up with serious intent then patients who for example suffer very real pain, such as in fibromyalgia, are likely to remain isolated from any real care. Even more seriously, biomedical research will be neglected.

7) An Unholy Alliance

In Manu's "the psychopathology of functional somatic symptoms" hidden away amongst the pages preceding the main volume is a quote from Hamlet:

> *Polonius.* I will be brief. Your noble son is mad.
> Mad call I it; for, to define true madness,
> What is 't but to be nothing else but mad?
> But let that go.
> *Queen.* More matter, with less art.

Manu makes clear within his preface his debt of gratitude to the Edinburgh psychiatrists Carson, Sharpe and later to Wessely and others. His remit is clear, that functional somatic symptoms are related to abnormal illness behaviour, sexual abuse, neuroticism, harm avoidance, sexual victimization and maladaptive coping. Such beliefs appear to be widespread as revealed both by Craig's survey which Manu also quotes, 83% of 400 physicians surveyed believed that patients with medically unexplained symptoms have a personality disorder. The words of Slater echo again *"In this penumbra we find patients who know themselves to be ill but, coming up against the blank faces of doctors who refuse to believe in the reality of their illness, proceed by way of emotional lability, overstatement and demands for attention."* It is indeed a common sense conclusion that if a patient has a physical complaint, sees a physician but gets nowhere, they will either seek help from another doctor or be more demanding as they return a week or so later. As for a patient with severe pain who does not constantly seek medical assistance when medical interventions fail, surely this is a more neurotic and abnormal form of behaviour? Yet when patients do present with medically unexplained symptoms, and especially when they are children, the consequences can be devastating. It is not unheard of for patients whose children present with so called pseudo-seizures to be told that the cause is, to quote "sexual abuse" and Lyme disease in children can be easily misdiagnosed as Munchausen syndrome by proxy[103]. The name of the neurologist in question here is withheld for obvious reasons. The conclusion must be that whilst some neurologists may well be "aetiologically neutral" in their use of the term "functional" the underlying currents within the profession are anything but. For a patient to expect an objective and unbiased assessment from Carson, Sharpe and other Wessely school psychiatrists is unlikely.

> *It remains the case that as many as one-third of the patients seen by clinical neurologists have symptoms that are better explained by neurosis than by neurological disease* [104]

The great difficulty with the current move to interpret illness in non-organic terms is the profound truth that everything is organic. Every atom in our bodies, every memory in our brains has an organic basis. The philosopher Heidegger called this our "being-in-the world." Within the context of this book the important point from this is that everything we are is a product of organic mechanisms. The reason you thought of a cream cake just now and maybe even actually tasted it is that seeing the word on the page a chemical reaction in your brain caused the brain to recall that memory. The problem is that even faced with gross abnormalities of function, or even abnormal behaviour within an individual, neurology often fails to suspect that there may well be an organic cause. In the same article Sharpe and Stone write[105]

> *The great psychiatrist Adolf Meyer, when shown the brain of a patient who had committed suicide at a postmortem examination, is reported to have challenged the pathologist to tell him by looking at the brain what was in his mind when he died. Important aspects of neurosis are likely to remain out of the reach of the scanner.*

If the pathologist had been more on the ball he might have pointed out that our species is the only one that chooses to end its life. He might have then pointed out that this is a genetic trait and that therefore certain individuals might be more prone to this genetic trait than others. At any rate, we are asked to make a value judgement, either what was in the patient's mind was an inability to cope with circumstances that would have overwhelmed any sane individual, in which case the reference to neurosis is absurd, or circumstances were actually slight in which case neurobiology rather than lesion localization would have been a more logical pursuit for the pathologist. That the pathologist, in the discourse presented, remains silent, speaks volumes for our fear of approaching the genetic basis of illness and indeed what it is to be a sentient human being. The psychiatrist Shorter writes:

> *The Nazi abuse of genetic concepts rendered any*

> *discussion of them inadmissible for many years after 1945. The notions of degeneration and inheritability became identical in the minds of the educated middle classes. Both were synonymous with Nazi evil. After World War II, any reference to the genetic transmission of psychiatric illness, whether as one factor among many or as inexorable degeneration, became taboo. The mere discussion of psychiatric genetics would, in civil middle-class dialogue, be ruled out of court for decades to come.*[106]

Given that functional paralysis is a mechanism present in all higher mammals and as Sacks[107] has noted, not in itself dependent on social and cultural mechanisms, Stone and colleagues would do well to explore the genetic issue further. If what in humans is referred to as "hysterical pregnancy" is common in howler monkeys[108] then to what extent are functional symptoms common in other animals? To what extent are genetic and therefore organic mechanisms responsible for the individual falling victim to disease that causes a disorder of function. There is some evidence that suggests that functional symptoms do run in families "Children who have family members with a history of conversion reactions are more likely to suffer from conversion disorder."[109] Yet rather than assume a hidden pathology the path espoused by Sharpe and colleagues is a return to 19th century notions of functional diseases of the nervous system and the implication that all functional symptoms have crucial sociological and psychological elements as a cause.

This is in marked contrast to purest approaches to neuroscience that attempt to liberate the individual from personal blame by uncovering the organic rather than personal source of their experience. The stressed single mother who develops functional weakness is likely to become even more stressed if her disability is seen as having social and psychological origins; she after all is the one left holding the baby. The chances are that being even less able to look after her child as a result of her illness, doctors who mention stress as a cause rather than an adjunct to biological factors, who focus on inappropriate coping mechanisms, are more likely to be seen as authoritarian representatives of the state than physicians wishing to help her and her child. We are thus left with exactly the situation Ssasz feared, as one psychiatrist, Dr. John Diamond states,

> *I am no longer a psychiatrist. I renounce it because I believe cruelty is at the core of the profession (and) I believe that there is something inherent in the profession that tends to bring out any cruelty lurking within. I have long wondered why this profession — which ought to be so compassionate – has, it seems to me, turned its back on humanity*[110]

It is with these words that we must examine the impact that psychiatry and an insistence on psychopathology as the singular cause of gross physical symptoms has had on individuals and families.

8) The case of Ean Proctor

For ten years I had been running a research unit specialising in chronic fatigue and the problems of people who are tired all the time.
- Prof. Simon Wessely[111]

In all M.E. epidemic or endemic patients the patients represent acute onset illnesses. The fatigue criteria listed here [in the CFS definitions] can be found in hundreds of chronic illnesses and clearly defines nothing.
-Dr Byron Hyde MD

In the autumn of 1986 a 12-year-old boy called Ean Proctor fell ill following a viral infection. He developed a range of symptoms including total exhaustion, feeling extremely ill, abdominal pain, persistent nausea, and recurrent sore throat, sensitivity to light and sound and loss of balance. He also began dragging his right leg, a monoplegic gait and thus a symptom often viewed as a classic "hysterical" sign. By 1987 his condition had rapidly deteriorated. Ean had gradually lost his speech and developed a paralysis that was to last 2 years. He was seen at the National Hospital in London by a neurologist, Dr. Morgan-Hughes. He reaffirmed the diagnosis of Myalgic Encephalomyelitis that Ean had received at home on the Isle of Man. In keeping with many experts on ME, Morgan-Hughes advised that exercise be avoided until muscle strength began to improve.[q]

During one visit to the National Hospital, a Dr. Simon Wessely approached the Proctors. Wessely was at the time a Senior Registrar in Psychiatry at the hospital. He asked Mr. and Mrs. Proctor if he could become involved in the case and help Ean. Faced with a severely disabled child the Proctors were desperate for help. Rather than admit that ME was a controversial illness and that he, Wessely, had particular views on its cause and treatment, Wessely inadvertently exploited the natural tendency of anyone who is desperate to accept any help whatsoever. Certainly at the time in Wessely's mind there was no question over the merit of the course of action he was about to undertake and one cannot rewrite history by using hindsight to suggest a lack of professional integrity.[r] Wessely like any doctor was keen to do what he could.

Wessely no doubt felt confident in his own clinical assessment of the situation and the assessment of two paediatricians supported this. Faced with a patient with no doubt functional somatic symptoms the kindest and most appropriate intervention was psychiatric. The problem of course is that one must set the sequela of viral infection not only within a contemporary but also within a historical context.

---

[q] In the acute stage of his illness the author himself found that vigorous exercise generally resulted in collapse with momentary loss of consciousness.

r See "The Life and Death of Harry Farr" for Wessely's own views on the use of hindsight.

As we have seen earlier Von Economo identified encephalitis lethargica as a *functional affection but on an organic basis*, what was missing in Wessely's approach, at least in the eyes of the Proctor's, was an acknowledgement that the "hysterical" signs afflicting Ean might also have stemmed not from Barbara Proctor's infliction of a "sick world" on her son but from a virus.

The cultural context within which Wessely was working was however one in which the very nature of viral infection had come under scrutiny. The fundamental notion that something as small and common as a virus could cause imperceptible damage yet gross dysfunction to the human body was at the core of this attack. Even more disturbing was the abandonment of the fundamental truth that Economo had dimly recognized, that disease might disturb the brain's notion of a limb or an arm with no doubt "functional" consequences.

Baring striking similiarities to the ridiculing of the Austrian physician Ignaz Semmelweiss, pioneer of handwashing to prevent infection, the early investigators of myalgic encephalomyelitis also found themselves undermined. Fundamental to this onslaught was McEvedy and Beard's work on the disease as mass hysteria.[112]

It is in this context that it is impossible to view Ean's case without reference to Akureyri, Iceland and the other ME epidemics. This is not simply because what happened at Akureyri in the 1940's informs our understanding of the disease that became known as ME but because McEvedy and Beard concluded this and indeed many if not all of the outbreaks were purely hysterical in which the patient's presentation bore no relation to any pathology:

> *These epidemics were psychosocial phenomena caused by one of two mechanisms, either mass hysteria on the part of the patients or altered medical perception of the community. We suggest that the name "myalgia nervosa" should be used for any future cases of functional disorder which present the same clinical picture. (McEvedy and Beard 1970)*

In 1947 Akureyri was hit by an outbreak of Myalgic Encephalomyelitis. McEvedy and Beard were at pains to point out that this epidemic "Akureyri Disease" as described by J Sigurjonsson was also a form of hysteria. Sigurjonsson noted the similarities between the condition he was faced with and that of polio. As a keen clinical observer he also noticed the differences, it was in effect an atypical poliomyelitis. The outbreak began shortly after the return to school for the autumn term and probably followed significant immunization as there had been polio outbreaks in neighbouring towns. The epidemic began among the school children before it spread to the adults. This was a feature of the way the disease spread that McEvedy and Beard curiously overlooked in their examination of a viral epidemic that took place in a school in the United Kingdom[113].

There were also similarities to an epidemic that took place in Los Angeles with both diffuse central and peripheral nervous system symptoms. Most notably when a poliomyelitis epidemic devastated the population of Iceland in 1967 the population of Akureyri remained curiously immune with blood analysis showing previous exposure to a polio type virus of the type alleged to have been responsible for the local ME epidemic in 1947. Hyde writes of the Iceland Epidemic:

> *Three children from this epidemic in the town of Friedrickshavn, became moribund and were unable to leave their beds, they eventually died of Parkinson's-like illness and were autopsied. Parkinson's Disease is almost unheard of in children. There can be no doubt that we were dealing with a diffuse inflammatory brain injury and at least some of these cases, involved the basal ganglia*[21]

It is perhaps likely in Ean Proctor's case that the work done by Wessely, Sharpe and other physicians influenced by McEvedy and Beard, rather than having a therapeutic input actually contributed to the complete social mess that the Proctors eventually found themselves in. The family clearly wanted pathology and wrenched from any historical context into the controversial mire of "chronic fatigue" they were simply not going to get what they desperately wanted. Was the pathology at Iceland an isolated case? Byron Hyde writes:

> *Circa 1996, an autopsy was performed on a woman with Myalgic Encephalomyelitis in Newcastle-upon-Tyne by Dr. John Richardson and the brain tissue examined by Dr. James Mowbray at St. Mary's Paddington. This woman had a history of typical Myalgic Encephalomyelitis, was well known by Dr. Richardson*

> *and accidentally died when her car fell of the side of the pier into the North Atlantic, the cold water preserving the brain tissue. Dr. Mowbray was able to demonstrate an autoimmune injury at that capillary level of the brain and basement membrane, the area that separates the capillaries from the neurons and brain tissue. In effect the same juxtaposition as in poliomyelitis but in this case in the brain and not in the spinal cord*[114]

Did Wessely think to check whether there was any vascular defect in the case of Ean Proctor? He could have done so without resort to autopsy. Peter Manu[115] in his discussion of SPECT (Single Photon Emission Computerised Tomography) and anomalies found in patients suffering from ME glosses over a seemingly crucial factor. Whilst patients with ME do indeed have similarities in perfusion to those of people suffering from Clinical Depression, there are also marked differences. Manu states:

> *Brain perfusion was assessed on three contiguous slices reconstructed for each region of interest, defined as $2cm^3$ of the basal ganglia, brain stem and cerebellum. Compared to patients with major depression, brain stem perfusion was significantly decreased in the chronic fatigue syndrome group, particularly those without psychiatric disorders.*[116]

The problem remains that repeatedly pathology turns up in the most unexpected places. In the case of Sophia Mirza a young lady who died in the United Kingdom whilst suffering from ME the neurologist Dr. Abhijit Chaudhuri writes that on autopsy he found:

> *"Unequivocal inflammatory changes affecting the special nerve cell collections (dorsal root ganglia) that are the gateways (or station) for all sensations going to brain through spinal cord. The changes of dorsal root ganglionitis seen in 75% of Sophia's spinal cord were very similar to that seen during active infection by herpes viruses (such as shingles)."*[17]

Prior to Mirza's death psychiatric intervention against a background of "abnormal illness beliefs" had resulted in her been sectioned, the police kicking the door down and her been forcibly removed from the family home.

Could it have been the case that establishing an organic basis for Sophia and Ean's illness would have allowed greater dignity for them and their families in their suffering? As Manu and Hyde make clear SPECT can be used as a diagnostic tool to differentiate between Myalgic Encephalomyelitis and a psychiatric diagnosis of depression or psychosis. Furthermore, autopsy has shown organic rather than "functional" causes for this alteration in blood flow in the brain.

Yet it seems that Wessely did not think that there could be any underlying organic pathology to explain Ean Proctor's presentation. Wessely soon informed the Parents that children do not get ME. On 3 June 1988 he wrote to the Principal Social Worker at Douglas, Isle of Man, a Mrs. Jean Manson:

> *Ean presented with a history of an inability to use any muscle group which amounted to a paraplegia, together with elective mutism. I did not perform a physical examination but was told that there was no evidence of any physical pathology...I was in no doubt that the primary problem was psychiatric (and) that his apparent illness was out of all proportion to the original cause. I feel that Ean's parents are very over involved in his care. I have considerable experience in the subject of 'myalgic encephalomyelitis' and am absolutely certain that it did not apply to Ean. I feel that Ean needs a long period of rehabilitation (which) will involve separation from his parents, providing an escape from his "ill" world. For this reason, I support the application made by your department for wardship.*

On 10 June 1988 Wessely provided another report on Ean Proctor for Messrs Simcocks & Co, solicitors acting on behalf of the Child Care Department on the Isle of Man. Interestingly Wessely had never once examined or attempted to communicate at length with Ean, he wrote, "I did not order any investigations....Ean cannot be suffering from any primary organic illness, be it myalgic encephalomyelitis or any other. Ean has a primary psychological illness causing him to become mute and immobile. Ean requires skilled rehabilitation to regain lost function. I therefore support the efforts being made to ensure Ean receives appropriate treatment." Under his signature, Wessely wrote "Approved under Section 12, Mental Health Act 1983."

Elective mutism is widely acknowledged as a response to stress in children, so Wessely could be forgiven for thinking that this was key to Ean's "hysterical" presentation. Yet Wessely clearly assumed that whatever stress was present in Ean's case was not the result of any organic pathology. Tests that did demonstrate organic pathology were, according to the family, ignored. Investigations performed by an ENT specialist in Manchester confirmed that Ean's vocal cords were not closing properly and that he was not exhibiting "elective mutism."

Whilst Chaudhuri's findings in the Mirza case show that organic pathology can make an individual sensitive to any stimulation with no doubt functional reactions, Wessely assumed that Ean's case was completely without organic basis "the prime problem was psychiatric."

In June 1988 social workers supported by psychiatrists and armed with a Court Order removed the child under police presence from his deeply traumatized parents. Ean was taken away and placed into care because of Wessely's "belief" that the illness was hysterical and was being maintained by an "over-protective mother." Communication between child and parents was censored and they were informed that they were not allowed to visit their son. To remove Ean from his "ill world" Wessely placed him within another "ill world" that of the psychiatric ward. Barbara Proctor writes:

> *We were eventually allowed to see the children's psychiatric ward that Ean was to be moved to. The door of the ward had a combination lock on it. One girl in the*

ward was literally ripping her room apart. She had both beds overturned, and things strewn all over the room. She was cowering in a corner screaming and swearing at the nurse. This scared the life out of us, and Rob and I knew immediately that this was no place for a child so ill with M.E., and were certain that they could not have treated children with M.E. in a ward like this.[118]

In Victorian times it was common practice to throw those suffering from what was then known as neurasthenia (a catch all term for anything from writer's cramp to chronic fatigue) into swimming pools. The belief was that the illness being "hysterical" would be resolved as primitive motor instincts came into play and the patient literally swam for their life. It is claimed[119] that Ean was therefore thrown into a swimming pool. It is also alleged that he was unable to save himself and sank to the bottom. Ean's parents claim that similar techniques were routinely employed to try and stimulate reflex actions and thus liberate Ean from the grip of hysteria. If it worked for the dog with supposed hysterical paralysis described by Oliver Sacks in "A Leg to Stand on" then surely it should work for Ean?

Stopping somewhat short of hosing him down in a manner typical of the Victorian asylum Ean was oddly taken on a ghost train ride. The hope was that he would cry out in fear and panic and this would prove the psychiatrists to be correct, Ean was suffering from the belief that he was sick. Irrespective of any illness this demonstrates more how out of touch the psychiatric profession were with the boy in their charge. The average eleven year old is unlikely to find a ghost train remotely frightening.

The Proctors claimed that Ean was also deprived of help to get to the toilet in the hope that he would avoid wetting himself and that consequently the boy was left for many hours in an urine soaked chair. Another supposed technique was to push Ean in his wheelchair at great speed up and down the corridors. The male nurse would stop without warning with the expectation being that the boy would hold on to the chair sides to stop himself from being thrown out. Barbara Proctor claims that he was unable to hold on and ended up on the floor. Was this false allegation, cruelty or distraction therapy? In re-iterating the Proctor's allegations to the House of Commons select committee Prof. Malcolm Hooper's 2003 briefing paper for the Countess of Mar pulled no punches.[120]

It would have perhaps been appropriate at this stage to question the diagnosis of Wessely, the fundamental assumption that there was no underlying organic pathology causing either the whole or part of Ean's condition. As we have already seen there is indeed a high co-morbidity of organic and "functional" illness. Yet from all accounts it remains unclear as to whether Wessely did the simple thing that every other doctor would have done in this case, examine the patient. No doubt the so called hysterical signs such as that of Ean's monoplegic gait were obvious and thus Wessely felt that further investigation could prove dangerous in affirming Ean's belief that he was ill. That such investigations could indeed have this damaging rather than liberating effect appears to be the belief into which Wessely and his colleagues were ordained. In contrast one expert on ME, Dr. Byron Hyde, writes of the great epidemics:

> *I have personally visited all of these cases except for the Cumberland epidemic and Wallis left us such a good description of that epidemic that there can be no doubt. I have personally gone to Los Angeles and examined patients from the Los Angeles epidemic. I have gone to Iceland and examined patients from the Akureyri epidemic. I have examined patients from the Royal Free Hospital epidemics, from the Newcastle sporadic illnesses. Many are the same or similar and many of them had been rejected or shunned because they were not true poliomyelitis. However they were all cases of Myalgic Encephalomyelitis*[121]

Not only did McEvedy and Beard fail to examine a single one of the patients from the Royal Free Epidemic, the Iceland Outbreak or those in Cumbria or Los Angeles, but Wessely also continued this tradition in dealing with Ean. He proceeded in the single mistaken belief that because no single organic cause for Ean's illness had been found, that there was unlikely to be one. When Ean failed to "snap out of it" psychiatry continued with even more gusto in its insistence that the child was effectively suffering from Munchausen's by Proxy and that the boy was deeply deeply traumatized, abused, every term that could be hurled at patient and parents. Fundamentally there was no technological reinvestigation of Ean. Hyde writes:

> *In addition to Brain SPECT, Brain PET, visual transcranial doppler to pick up CNS arterial spasm disease as well as obstruction, QEEG is essential for the physiological assessment of brain dysfunction.*[122]

Yet no such technological investigation was offered to Ean. As one of Wessely's colleagues Michael Sharpe writes:

> *This leads me onto my final point which is the difficulty providing effective treatment for patients in the real world. The reality is, there is almost no availability of specialist Cognitive Behavioural Therapy for patients with Chronic Fatigue Syndrome or any of the other related unexplained somatic syndromes in Scotland, why? The reason is of course due to NHS Priorities for use of resources. But given that the therapy is relatively cheap, it is also related to attitudes.*
>
> *Purchasers and Health Care providers with hard pressed budgets are understandably reluctant to spend money on patients who are not going to die and for whom there is controversy about the "reality" of their condition.*[123]

It is therefore the case that any treatment patients must enter into when the "reality" of their condition is controversial, is one in which the psychiatric paradigm is perpetuated. Sharpe argues strongly for more treatment for these patients. It is however treatment in the form of CBT (cognitive behavioural therapy) which 93%[124] of severe ME patients greet with the same gusto as a badger does the notion of being gassed. It is an erroneous idea to believe that patients, unlike badgers can run away. The sick and the vulnerable regrettably need doctors, even if the treatments meted out have little benefit, or at worst may, as Richardson, Hyde and Cheney suggest, lead to cardiac arrest.[125] Such a desire to treat symptoms rather than the actual cause has the same reality as having the patient pretend they are climbing a mountain rather than actually climbing one. In the author's case such CBT bravado led to collapse whilst climbing at altitude above basecamp in the Tien Shan. The ensuing rescue cost $20,000.

Ean was therefore caught between the possible "way out" that an organic explanation for his condition would have offered him and his parents and the ideological refusal to grant them one by the psychiatrists. Such an organic explanation would have been both socially, scientifically and politically wise and expedient. It would have allowed a neutral approach to the situation.

In ME the blood flow to the brainstem is severely impaired. The part of the brain that mediates all aspects of the body's function is thus heavily compromised.[126] If any part of Ean's illness was indeed an emotional reaction to the horrors that had overtaken him then that could have allowed a role for psychiatry. His parents and the neurologist who first diagnosed their child would not have lost face and Ean's case could have been added to the long scientific narrative that traces its way through the many outbreaks of ME right back to those first described by Wallis in the Cumbrian outbreak. Instead Wessely wrote on the 5th August 1988, directly contradicting the neurologist Dr. Morgan-Hughes:

> *A label does not matter so long as the correct treatment is instituted. It may assist the Court to point out that I am the co-author of several scientific papers concerning the topic of "ME"....I have considerable experience of both (it) and child and adult psychiatry (and) submit that*

> *mutism cannot occur (in ME). I disagree that active rehabilitation should wait until recovery has taken place, and submit that recovery will not occur until such rehabilitation has commenced.....it may help the Court to emphasise that...active management, which takes both a physical and psychological approach, is the most successful treatment available. It is now in everyone's interests that rehabilitation proceeds as quickly as possible. I am sure that everyone, including Ean, is now anxious for a way out of this dilemma with dignity.*

Sharpe in his book "Chronic Fatigue Syndrome ME/CFS the facts" makes clear his approval of psychiatric management techniques. It matters not one jot that people become fatigued for a variety of reasons, nor even that a diagnosis of ME is based on central and peripheral nervous system problems exacerbated by exercise. For Wessely and Sharpe the use of "Chronic Fatigue" is a criteria as broad as used by Weir Mitchell and the Victorian counterparts they seemingly admire. Put simply there is no such thing as ME. Like all other illnesses affecting the function of the central nervous system it is a neurosis, a modern day term for neurasthenia which can be resolved through nerve tonics and the protestant work ethic of "take up thy bed and walk"[127]

The net result of the Proctor case was that it became a rallying point for the ME community against the psychiatric profession and perhaps unfairly against Wessely himself. Wessely himself has thus been forced to defend his actions and that of the paediatricians involved in Ean's care. Even though he continues in his belief that

Ean was suffering from hysterical symptoms compounded by his mother's illness beliefs, such interventions as removing the child are, he now claims, not appropriate; "You just can't do that kind of thing"[s]. Asked by Sheena McDonald in a Channel 4 News Programme on 26th August 1998 as to whether taking a child from its parents could be justified, Wessely curiously stated, *"I think it's so rare. I mean, it's never happened to me."*

The Tynwald Select Committee's report that followed in the wake of Ean's case, found against the medical profession and in favour of the parents and it recommended that another enquiry be set up to decide compensation. However, the McManus Enquiry set up to examine this issue found that the Proctors were not entitled to anything and that they were responsible for Ean's illness. This finding, in opposition to the Tynwald Select Committee Report effectively helped undermine ME as a neurological condition in the UK and plucked victory out of defeat for the psychiatric lobby.

Ten years later Mrs. Proctor answered her door on the Isle of Man and was surprised to find herself facing a police officer who had been directed to question her by the Metropolitan Police. Mrs. Proctor's name and address had been found in the house of parents who had found themselves in exactly the same position as the Proctors.

Fearing the same fate as Ean however child X's parents had taken their son abroad. When the father returned he was arrested and sentenced to 2 year's imprisonment. The police assumed that the

---

[s] *From a private source*

Proctors must be harbouring child X, despite the fact that the family had even received the support of her Majesty the Queen in a letter acknowledging the plight of their son in suffering this "disease".[128]

Illness does not exist in isolation, it occurs in individuals who will all react differently both physically and emotionally. It takes place in communities who will all place their interpretation on what is wrong with the patient. The expectations that a community forms and places upon an illness will have a reciprocal impact on the presentation of the patient.

This does not mean that there is no disease; rather it is a self-evident description of all illness throughout history. That in primitive society disease is associated with demons and curses does not mean there is no pathology. It may well mean a very different presentation from the person who believes they have been cursed. Similarly doctors themselves must proceed in a manner that fits with the patients and societies own beliefs. In India, for example, sneezing is considered to be a sign of good health. In a technological age when technology has failed to provide a narrative to explain the patient's condition the modern priests and shamans, the psychiatrists, are even more likely to find themselves met with rejection. In the 21st Century medicine that describes any symptoms or signs, especially physical symptoms and signs, through the use of terms such as neurosis has fundamentally alienated itself from the society within which it operates.

Faced with symptoms that cannot be explained by tests, psychiatry has constantly resorted to stories of somatisation. That what it views as physically unexplained symptoms are in fact emotional disturbances. In writing to the e-British Medical Journal in a letter dated 29th December 2003, Angela Kennedy a Social Science Lecturer states:

> *I suspect that psychiatry, if it is not careful, will eventually become most ridiculed over its adherence to one theme: that of 'somatisation'. Presently, sufferers of Myalgic Encephalitis (sic) (also called Chronic Fatigue Syndrome) are increasingly subject to medical negligence or even abuse because the huge body of international bio-medical evidence is ignored, especially in Britain, in favour of an unfortunately incomprehensible, incoherent and empirically inadequate theory.*
>
> *"The categorization of an illness as being psychosomatic also means a further categorisation of an individual as 'deviant' rather than 'ill', so that they are denied sympathy, support, and even benefits they are entitled to. Categorised as 'deviant', the ill then suffer increasing social exclusion and material inequalities.*
>
> *"The main problem with somatisation theories is that they cannot be either proven or disproven and therefore are not very 'scientific' at all. (Kennedy 2003)*[129]

Yet as we have noted, when tests are negative patients who are deemed to have "functional" symptoms are indeed viewed as deviant. They are a challenge to the hegemony of medicine; they cannot be medically explained in one simple definition, as can "classic" illnesses such as Multiple Sclerosis or Parkinson's. They are also likely to use up huge resources as they struggle to find a diagnosis with which they are happy. The right to have tests may be seen as a basic human right in a democracy but it is not so in medicine[30]. Patients are therefore not deviant within the society from which they come but deviant within the society they have entered into relationship with.

Doctors do not display their corpses as a reminder of past mistakes or readily admit to ignorance or error, nor do they like their budgets spent by people who question their authority and thus obviously have a "personality disorder"[130].

To avoid a crisis of confidence in both itself and the profession as a whole, neurology must "sell the diagnosis." As Stone and colleagues' research has shown and as has been noted throughout this book, the term most effectively used to sell a diagnosis of hysteria is the term "functional", a word that both Sharpe and Stone see as rooted in the pre-Freudian approach to patients with medically unexplained symptoms. Interestingly it has been claimed by Stone and colleagues[131] that the millions spent on MRI and CT scanners have made little difference in reducing missed diagnoses. In "A systematic review of misdiagnosis of conversion symptoms and hysteria," they state, "The five yearly misdiagnosis rate fell to 4.4% (2.1% to 9.2%) in the period 1970-4, which is *before* computerised tomography became generally available" (emphasis added). This discovered rate of misdiagnosis has remained roughly at 4% to the present day.

It is of course impossible to define how many patients have missed organic anomalies that could explain their condition. This is due to the fact that, if missed at onset, only rapid deterioration or death and autopsy are likely to reveal them. Stone and colleagues write that in the use of MRI and EEG "Our data suggest that they have not been as important as previously believed."

Remarkably conditions which were often previously diagnosed on clinical features and history, conditions such as multiple sclerosis, have 8% of patients reclassified as suffering from neuroses when the scanner fails to show plaques of sufficient size and quantity. As the diagnosis of MS and the interpretation of MRI in this regard is itself a controversial issue, is it not more likely that a substantial number of patients have organic disease beyond the limits of the scanner? In relation to ME which Sharpe and colleagues view as "functional" the irony is that in what Hyde defines as a vascular condition it is precisely those tests that are still not routine which excel at diagnosis, SPECT, Doppler etc, MRI is in Hyde's own words little better than an x-ray. In the ME outbreak in Cumbria it was only through autopsy Wallis found that:

> *There are in the entire diencephalon, particularly around the third ventricle, numerous small haemorrhages, which extend into the adjacent parts of the mid-brain. Similar haemorrhages can be seen in the corpora mamillare. The haemorrhages are mostly around the small vessels but some are also to be seen in the free tissue. This is a significant finding.*[132]

9) The age of Steam

You have managed to write a book about hysteria without even mentioning the case of Anna O, the most famous 'hysterical' patient of them all- Richard Webster

Who is Anna O? – Prof. Simon Wessely

With industrialization in Europe the old ties of community and local responsibility for the sick began to fragment. For the mentally ill this meant the growth of the asylum where previously those deemed to be psychotic, neurotic, socially deviant or simply difficult to manage were looked after in their own villages, often in appalling conditions. It was not uncommon for the sick relative to be kept shackled in a barn or confined amongst the animals. So in many ways, though in no way ideal, the asylums and growth of hospital care for the mentally ill was a step forward. The fragmentation of care away from communities and towards corporate and centralized management continued right through to the end of the 20th Century.

In Britain this meant the gradual closure of local "cottage" hospitals and the delegation of out of hours services away from a patient's own GP to that of specialist "emergency" doctor services. Increasingly patients were treated in large "institutionalized" establishments that were themselves assessed against strict performance criteria. Financial concerns became more important as it was realized that taxpayers were unwilling to pay for a health service that was increasingly short of resources.

For those deemed to be suffering from medically unexplained symptoms the removal of the physician from the community was a disaster. The Newcastle based GP John Richardson describes visiting a patient at night who complained of suffering palpitations and heart problems following a viral infection. He had also documented in the community the association of myocarditis with the same strain of enterovirus that had affected the woman and her family. Richardson struggled however to get colleagues really interested in the cardiology of the patient. Richardson writes:

> *In addition, some thought that she was making heavy weather of her muscle pain and thought that she was psychologically depressed.*[133]

On a bleak Tuesday morning in November 1989 Richardson was called by the woman's husband to the family home. The woman was distressed that she might have wasted her doctor's time but Richardson told her that she should not think like that and it was always a privilege to attend her and try to help. She then had a sudden collapse and died due to a dysrhythmic cardiac arrest.

Had the same woman presented following her viral infection in a contemporary Accident and Emergency Clinic in the weeks prior to her death there can be little doubt that isolated from family history and the insight of a physician who had attended the whole family throughout their lives, the diagnosis would have been different. Her notes would reflect the fact that she had presented in Accident and Emergency with symptoms that specialists felt due to depression. No doubt her notes would also state that examination had failed to find any pathology sufficient to explain her symptoms and that there was a feeling that her abnormal heart rhythm was caused by stress. The assumption would be that her presentation was psychosomatic.

Crimlisk and others make clear that trawling through a patient's notes for such incidents is crucial to the diagnosis of hysteria. So that if the woman had presented again, outside the context of family and community, anything atypical about her presentation would be seen as a sign of somatisation, conversion disorder or whatever term the physician chose to use. As we have already seen Crimlisk and others suggest that in such cases further tests and investigation are dangerous to the health of the patient as they may well re-enforce illness beliefs. A heart complaint may not necessarily result in death, it may result in enormous distress and emotional trauma for the patient as they suffer non-organic explanations, offers of psychotherapy or CBT. Strikingly if the road suggested by Crimlisk and others is followed to its natural conclusion the chance of actually finding any organic pathology is more and more unlikely. "Furthermore, because you do not find pathology does not mean there is none" (Byron Hyde).

Whereas Richardson may have remembered that the woman's child also had a slight heart murmur when first born and that this had worsened during a viral infection, such insights are beyond the scope of the ultra-specialist in the county hospital who is perhaps even unaware that the woman has children. In order to gain the expertise of insight into specific medical conditions a sacrifice is made. What is lost is the realization that disease does not occur in isolated individuals but is part and parcel of community and family life. It can be both subtle and overt in presentation. The growth of ultra-specialisation within the NHS has meant that it is the acute and obvious illness that is most likely to be diagnosed.

With the growth of urbanization at the start of the 19th Century a similar dilemma was faced. It was made even more acute by the fact that x-ray, MRI, even the lumbar puncture were all inventions of the future. Faced with a growing population of "the sick" management became more and more of an issue. This became even more apparent where patients did not fit neatly into classical disease definitions. Terms that originally had meant quite specific things, hysteria a disease primarily affecting women and associated with the womb, neurasthenia a vague disorder of functioning that could include anything from headaches to fatigue, all became rapidly enlarged to include any patient who could not be given a "classical diagnosis."

It was in this environment that the French neurologist Charcot began to develop his ideas about hysteria as a disorder of the nerves. Concepts that had little experimental basis and that he himself broadened to such a degree that they were challenged by his successors at the Salpetriere. Whereas in the past hysteria had been classed as psychogenic, deviant or simply insane, Charcot's concept of hysteria as a disorder of nervous functioning allowed hysteria to be taken out of the asylum and used in mainstream medicine.

Historically the ideas of Freud in relation to hysteria remain vitally important, but it is the contribution of Charcot, Janet and Babinski that are of most relevance to contemporary neurology. Chris Bass comments "Freudian theory dominated thinking from the turn of the twentieth century. Freud considered the symptoms to be a result of a conflict that the patient could not resolve: this conflict was then "repressed" or thrust into the "unconscious" and the resultant symptom was the only sign that intrapsychic battle had ended. Needless to say there is no evidence to support these views."[134]

Freud argued that developmental disorders, sexual abuse in childhood, all could lead to unconscious symptom formation via the mechanism of repression. Indeed Martin L. Gross explained Freud's own migraines by suggesting that he had seen his mother having sex when younger and that this repressed memory was the cause (Martin. L. Gross, "The Shadow of Freud").

As we have already noted, contemporary neurology continues to explain epileptic seizures that do not register on EEG as "most likely caused by sexual abuse" (name of consultant withheld). This is despite the fact that EEG only measures the electrical activity on the first few millimetres of the skull with even skull and scalp thickness producing variations[135]. As Bass also states "Modern definitions continue to regard the symptom as arising from emotional conflict: it is assumed that the patient's ability to exercise a conscious and selective control is impaired,"[40].

To understand how this "assumption" that affects 40% of neurological patients came about we need to look back to one of Charcot's own patients, a man named Le Log. Richard Webster writing in "Why Freud was Wrong" states:

Le Log--- was a florist's delivery man in Paris. One

evening, in October 1885, he was wheeling his barrow home through busy streets when it was hit from the side by a carriage which was being driven at great speed. Le Log---, who had been holding the handles of his barrow tightly, was spun through the air and landed on the ground. He was picked up completely unconscious. He was then taken to the nearby Beaujon hospital where he remained unconscious for five or six days. Six months later he was transferred to La Salpêtrière. By this time the lower extremities of his body were almost completely paralysed, there was a twitching or tremor in the corner of his mouth, he had a permanent headache and there were 'blank spaces in the tablet of his memory'. In particular he could not remember the accident itself. But, because there had never been any signs of external injury, Charcot decided that Le Log--- was a victim of traumatic hysteria and that his symptoms had arisen as a result of the psychological trauma he had suffered. Charcot came to this conclusion knowing full well that some weeks after his accident Le Log--- had suffered heavy nose-bleeds and a series of violent seizures – seizures which Charcot deemed hysterical.

In the century which has passed since Charcot made this diagnosis, the face of neurology – and of general medicine – has been transformed. If Le Log--- were to be brought today to a hospital in practically any part of the Western world there can be no doubt that doctors would recognise a case of closed head injury complicated by late epilepsy and raised intracranial pressure.

From this we may derive a conclusion which is both simple and terrible in its implications: Le Log---, the classic example of a patient who supposedly suffered from traumatic hysteria, did not forget because he was frightened. He forgot because he was concussed. His various symptoms were not produced by an unconscious idea. They were the result of brain damage.

At the start of the 21st Century we have become more and more aware of how seemingly non-organic conditions are in fact rooted more in neurology than psychiatry. Indeed psychiatry as a discipline has seen a dramatic fall in medical students willing to specialize in an area that is increasingly seen as less than cutting edge.

Following regular attendance at case conferences over a number of years one health service manager remarked "psychiatric diagnoses are randomly plucked out of the air, often even after several weeks of discussion the basic fact as to whether a patient is depressed or euphoric cannot be decided"[136] In the 1990's that bedrock of psychiatry, schizophrenia, eventually yielded to organic investigation. By 1995 the gene or genes causing schizophrenia had been placed roughly somewhere on chromosome 6. For manic-depressive illness, chromosome numbers 18 and 21 had been identified.[137] With regard to Schizophrenia, imaging techniques also began to shed light on what had so often been seen as a developmental disorder. A disorder that the Scottish psychiatrist R. D. Laing believed originated in the community and very family from which the schizoid personality emerged. Laing failed to realize that whilst the content of the patient's presentation may indeed have been forged from the community that they came from, the underlying brain disorder was not. Yet again we are reminded of Osler and the epigram to this book: *no individual reacts or behaves alike under the abnormal circumstances we call disease.*

At the same time as psychiatric conditions have increasingly been seen in organic terms there has been growing discontent amongst patients to both accept and respect diagnoses for physical symptoms that suggest there is no underlying pathology. At the start of the 20th Century Hurst was able to use his authority as a physician and status in society to cure those patients who presented to him with functional symptoms. Some of this may as Webster suggests have been due to the placebo effect[138] but Hurst's ideas of active physiotherapy and rehabilitation also played a part. He was not faced with patient advocacy groups such as "the one click group" demanding that he explain the presentation of his patients in organic terms and encouraging patients to reject what he had to say.

Contemporary neurology therefore faces a profound difficulty in managing patients for whom there is no current organic explanation despite claims of "extensive" but also appropriate investigation. The notion of functional disorders of the nerves thus becomes much more attractive just as at the end of the 19th Century the middle classes rejected the treatment of the asylum and classifications that insinuated there was something wrong with them as a person.

Increasingly 20th Century neurology and psychiatry has attempted to hold on to its authority by adopting this middle ground between organic and psychogenic causes for unexplained symptoms in a bid for acceptability. Indeed studies of patients given a diagnosis of "functional" show that they did remarkably well in terms of recovery right through the 19th Century and into the 20th. By the 1930's however neurasthenia and other functional disorders were increasingly seen by neurologists in psychiatric terms. By 1932 they had vanished altogether into a large category of psychoneurosis, terms such as "maladjustment" and "hysteria" that remain unpalatable and un-therapeutic for most patients.[139] It is little wonder that some should seek to rescue the term "functional" from this later semantic shift.

Yet with the increasing claims of neuroscience this shift is likely to be towards organic explanations, it is likely to continue and indeed the entire history of medicine suggests this will be the case. In adopting this middle ground, this "third way", the notion that a reciprocal interaction between the mind and the body can both cause and be used to treat symptoms is central to both philosophy and clinical practice. Yet to confuse the work of those who seek to occupy this "third way" the term is often still reduced in the NHS to simply a psychiatric condition. Functional weakness thus becomes a diagnosis in which the "psychology in these cases is complex and uncertain" as recorded in patient notes by one neurologist, a Dr. David Bateman.

The idea of functional nervous disorders has its roots in an age of industrialization and the growth of the middle class, an age in which concepts of the self were based more on judaeo-christian notions of the mind body and spirit. As a society we have shed the notion of spirit from this tripartite description of what it is to be a human being, it is likely that the notion of mind as a distinct entity will through neuroscience be removed also. Some neurologists will no doubt continue to cling to the idea that the mind can have an impact on the body and that the ideas occupying a person's thoughts, illness beliefs or otherwise are somehow separate from their biological makeup. If, as Dennett suggests, consciousness is merely the froth on the cappuccino of a vastly complex organic mechanism then privileging its ability to create not just symptoms but the stigmata of hysteria requires enormous faith. Human beings are organic machines, different only from amoeba and tigers by degrees of complexity.

Historically the idea of functional nervous disorders was far from a distinct classification of illness. Rejecting the horrors and stigma of the asylum the affluent middle classes sought help from an increasing number of "nerve doctors." Treatments were diverse from the talking cures offered by Freud and his associates to the growth of spa therapies, taking the waters and other alternative cures. These were all supposed to restore an imbalance in nervous functioning.

Neurasthenia, mentioned earlier, came to be a popular diagnosis, a term coined by George Beard that described a vague condition with symptoms of fatigue, anxiety, headache, impotence, neuralgia and depression. Contemporaries enlarged this further with even symptoms of weakness, repetitive strain injury and fainting. Women were thought to be particularly prone to the condition and rest was advocated with the expectation that this would be total and prolonged. The Rexall drug company introduced a medication called

"Americanitis Elixir" which it claimed to be an effective treatment for any bouts related to neurasthenia. In effect it was the prozac of its day.

Today, the term neurasthenia is viewed by some as an attempt to group together a wide variety of different and poorly understood disease processes, chronic fatigue, fibromyalgia and dysfunction of the autonomic nervous system, dysautonomia.[140]

The lessons learnt from the study of disorders of autonomic function, weakness caused by postural hypotension or postural orthostatic tachycardia (POTS) with its associated bladder dysfunction is that statements such as "your central nervous system is not broken it's just not working properly" effectively do not mean very much.

The nervous system may well be intact but it is only one part of an organism where neurotransmitters and dendrites must work in harmony. To write on patient notes "unlikely to be organic in origin" is an incredible act of either bravery or downright historical naivety.

The dilemma increasingly faced by the medical profession is that though tests are indeed often negative, patients routinely reject psychiatric explanations, somatoform autonomic dysfunction, psychogenic weakness and conversion disorder diagnoses. The bedrock of psychiatric diagnosis, the DSM handbook, offers no scope for the emerging realization that disease is classified by pathology, not by classification. The entry for somatisation disorder states:

Somatisation disorder: History of many physical complaints; 4 pain sites or functions: 2 nonpain gastrointestinal, 1 sexual or reproductive, 1 pseudoneurologic. Onset before 30 years of age and not explained by general medical condition or substance effect[141]

The immediate objection to this is that the patient cannot be suffering from somatisation disorder if the fourth pain site accompanied by a pseudoneurologic symptom happens the day after their thirtieth birthday. For women, "reproductive" could mean an inability to conceive and equally the poor quality of a man's sperm could, on strict appliance of the criteria, be deemed to be caused by "worry" if no other explanation can be found as to why no sperm should be able to penetrate the female egg. Such tick list definitions of disease including the Fukuda and Oxford criteria for Chronic Fatigue Syndrome are thus diagnostic madness, relating as they do to a bureaucratic need to classify rather than to diagnose and treat pathology.

The criteria for conversion disorder is equally absurd, with secondary gain deemed to be a key feature. Based on this every child who has ever eaten ice cream after the removal of its tonsils has already fulfilled one criterion. As not being able to talk properly is a common complication of tonsillectomy then impairment of social functioning means they have fulfilled another criterion whilst if the reason as to why they have had repeated throat problems and long lasting throat infections remains a mystery then another tick can be made against the diagnosis. Given that the most severe infection occurred at exam time with associated distress we are already well on the way to a diagnosis of conversion disorder. The surgeon is therefore guilty not of relieving suffering but of iatrogenic harm. The DSM states:

Diagnostic criteria for conversion disorder DSM-IV 300.11 are as follows:

- *One or more symptoms or deficits affecting voluntary motor or sensory function suggest a neurologic or other general medical condition.*
- *Psychological factors are judged to be associated with symptom or deficit because initiation or exacerbation of symptom or deficit is preceded by conflicts or other stressors.*
- *The symptom or deficit is not intentionally produced or feigned (as in factitious disorder or malingering).*
- *The symptom or deficit cannot, after appropriate investigation, be fully explained by a general medical condition or by the direct effects of a substance or as a culturally sanctioned behavior or experience.*
- *The symptom or deficit causes clinically significant distress or impairment in social, occupational, or other important areas of functioning or warrants medical evaluation.*
- *The symptom or deficit is not limited to pain or sexual dysfunction, does not occur exclusively during the course of somatization disorder, and is not better accounted for by another mental disorder.*

In the United Kingdom there has been a growing backlash against DSM with diagnosis based not on psychological factors but the absence of organic pathology and the presence of the so called physical "signs" of hysteria, the stigmata that were identified initially by Charcot Janet and Babinski.

As we have already seen however alongside those with extensive organic brain injury it is homosexuals, women, the insane and intellectuals who are most likely to present with these symptoms and be misdiagnosed as a result. That homosexuality only narrowly escaped entry in DSM III as a deviant sexual behaviour must make us reflect that the stigmata of neuroses as objective "signs" are not entirely free from cultural interpolation.

In his article "Somatoform disorders: a help or hindrance to good patient care" Sharpe and Mayou state that the use of somatoform disorders as a diagnosis where symptoms are unexplained by disease "has been unhelpful by perpetuating dualism." They see functional symptoms as on a continuum between organic and psychiatric diagnoses. Sharpe and Mayou state that:

> *Medicine can only continue to assume that that all their patients' illnesses are accounted for by disease pathology if they can label those patients whose somatic complaints do not fit this assumption as really 'psychiatric'. Psychiatry can only continue in its belief that these somatic complaints are really based purely in psychopathology by labeling them as having somatoform disorders with the associated and dubious implications of 'somatisation'*[142]

Whilst usefully highlighting the absurdity of an either/or diagnosis, that essentially if tests are negative medicine tends to describe bodily complaints with psychiatric terms, Sharpe and Mayou fail to understand what is commonly understood by the public to be a disease:

Disease: a disorder of structure or function in a human especially one that produces specific signs or symptoms:
OED (2007)

In this sense all psychiatric diagnoses are essentially diseases and organic. Depression is associated with serotonin and those disorders that have psychological factors can be explained by underlying pathology. Multiple sclerosis, Parkinson's- even the common cold is seldom associated with a soaring of spirits.

This is not to say that patients do not present to doctors because they are sad or troubled but rather that these are life events, not illness, and medicine has become inextricably tangled in this mire when its remit has always been to treat disease.

There has of course been significant research into the use of the term "functional." In "What should we say to patients with symptoms unexplained by disease" Stone et al surveyed the response of patients to a number of diagnostic terms[143]. Multiple Sclerosis and Stroke came highest in terms of acceptability with the term "functional" a close third. Patients were not told however where the term "functional" originated from, nor its relationship to one of the least acceptable terms "hysterical weakness." 52% of patients were offended by the term "hysterical weakness" and 93% by "symptoms are all in the mind." What is profoundly unclear from this research is the extent to which patients understood the terms that were put to them. It could equally be argued that offence was based purely on the pragmatic context and understanding within which researcher and patient operated.

Patients who had been primed with the aetiology and semantic meaning of the terms they were presented with may well have had a very different opinion on whether a term was acceptable or not. If they understood that for many doctors "functional" means that the patient is neurotic then it is highly unlikely an informed patient would find the term any more acceptable than "hysterical weakness."

Irrespective of this, the researchers fail to appreciate that it is symptoms that are explained by disease that have the highest acceptability. The cultural gap between researchers and patients could not be wider as the very title of their article makes clear "What should we say to patients with symptoms unexplained by *disease?*" (italics added). The pretext is clearly that they are not suffering from disease and the sooner they can be made to accept that their symptoms are caused by the reciprocal interaction of "mind" and body then the more likely they are to get better.

The risk is that such terminology may well in the patient's mind at least be associated with disease. Its use can thus become a slight of hand to prevent the patient rejecting the diagnosis as indeed would 52% if they were told their symptoms were "hysterical." It would be wrong to assert that this "slight of hand" is a deliberate attempt to deceive; the "trick" of acceptability comes from a misunderstanding on both sides. 83% of patients felt that "functional weakness" was a good reason to be off work, third only to Stroke and Multiple Sclerosis.

Functional is therefore a term that is not accepted because it is neutral, but rather because it carries the mystique and suggestion of real illness as indeed one doctor clearly found out:

> *A professor of medicine was consulted by a middle-aged lady with stomach ache. On her second visit he looked one by one at her x-rays and then turned to her, beaming, and said, 'I'm glad to be able to tell you that they don't show anything wrong.' Seeing her crestfallen face, he asked her if that was not what she wanted to hear. 'Well, my doctor told me,' she exclaimed indignantly, 'that I've got a large functional element.'* [144]

The extent to which descriptions of disease offend is thus only partially understood within the context of medicine; it is within the context of society that these terms gain their full meaning. If patients do not understand the terms that are used to describe their condition then the physician is not performing a role of reassurance and explanation but one of manipulation and control.

Rather than call for an abolition of dualism as unhelpful for patients, Sharpe and Mayou could perhaps consider an abolition of non-organic diagnoses. The patient who is off work due to a problem with serotonin is likely to find life easier within society and the work place than the patient who is "depressed."

10) Conclusion

Hysteria as a clinical entity relies on the notion that the nervous system must be intact in the context of apparent neurological dysfunction. By implying mysterious unconscious mechanisms and the notion of hysteria as an emotional disturbance, its proponents have created a system of infinite regression. To quote Stephen Hawking's satirical comment on the origins of the universe, with hysteria it is also, "Turtles all the way down". That is to say mechanisms of "hemi-depersonalisation" or "dissociation" are not explained as proceeding from biochemistry but rather themselves proceed from psychological states. So that one is left with "hysteria all the way down". Hysterical symptoms are caused by, well, hysteria. Concepts of "the will" rather than autonomic function prevail. Complexity in human beings is seen as cognition rather than the chaotic response to sympathetic or parasympathetic stimuli.

As we have seen social and emotional aspects of hysteria cannot be seen as diagnostic tools due to their prevalence in all illness. Furthermore, the high percentage of co-existing organic pathology and often a past history of neurological disease makes the use of the term "conversion disorder" effectively meaningless. It is simply not that easy to work out that the nervous system is actually intact and that there is a complete absence of contributory organic pathology. Even where it is claimed fMRI has shown an apparent link between symptoms of weakness and florid emotional trauma, the persistence of that weakness relies on defective vasomotor mechanisms rather than the Freudian unconscious.[145]

The difficulty posed by hysteria has not receded with time; the difficulties that began to be recognized in the 19th Century remain. In fact the very naming of the condition troubled Babinski who felt it should be called Pithiasm due to the alleged ability to cure hysteria through suggestion (Peitho- I persuade and iatos- curable). This contrasted with Janet who felt that the term first suggested by Charcot should be preserved. As we have seen, the neurologist Hurst used a kind of internalized physiotherapy to reconnect circuits that had ceased "talking" to each other following organic injury. The debate over what to call "the diagnosis that dare not speak its name" continues today. Such debates are far more than semantic, they have profound implications for how patients are diagnosed and who shall treat them.

The Freudian notion of the unconscious that has held sway over much of western thought originated from judaeo-christian notions of sin and disorders of volition as morally deviant. Such notions have contaminated our western thinking about the nature of the body and mind so that psychiatrists are left to treat the "myriad of diseases yet to be recognized." As neuroscience continues to push back the boundaries of our understanding of the brain, psycho-social explanations for what are increasingly understood as organic disorders have less and less relevance. Writing in "A neuroscience of hysteria" Broome states:

> *For hysteria to survive neuroscientific enquiry our psychological intuitions must be realizable neurologically. If this is not possible then either the traditional conceptualization of hysteria or our hope for a scientific*

> *psychopathology would require revision. Such concerns do not just impact upon hysteria, but ultimately influence our conceptualization of mental illness and the psychiatric taxonomy employed.*[146]

Contemporary concepts of functional symptoms recognize that a dualistic approach is far from satisfactory yet the final push of the gates through to a purely organic explanation has yet to be made. The resort to 19th Century "disorders of the nerves" as an attempt to rescue a tradition that existed before Freud can never wholly extract itself from 19th Century thinking and beliefs. This after all was an age in which office psychiatry grew out of a middle class preference for it rather than the asylum and in which Babinski believed hysteria could be prevented by hygiene. Charcot himself believed that there was a connection between hysteria and magnetism, whilst Janet favoured explanations based on hypnosis. Such a menagerie of explanations continues to haunt studies of hysteria that have become increasingly bogged down in symptoms research rather than evidence based laboratory medicine. The call to abolish the many categories of somatisation from the DSM, although logical, calls up the spectre of defining psychosomatic disease in the absence of psychological symptoms. Simon Wessely writes:

> *The suggested distinctions that appear to have some empirical and practical validation would be a diagnosis that continues to insist that either symptoms and/or loss of function be inexplicable in conventional biomedical terms and then distinguish between symptoms and loss of function, and between acute and chronic onset of either.*[147]

Wessely himself applied such explanations to Gulf War syndrome. Faced with symptoms that seemed inexplicable in conventional biomedical terms he favoured a psychological explanation of the condition. Wessely felt that the only thing that could have affected so many different people was stress, especially anxiety about chemical weapons. This in itself was odd as other studies showed the cluster of symptoms experienced by Gulf War veterans to be unique to the first Gulf War. Wessely also felt that misinformation about Gulf War syndrome afterwards and fear concerning the many vaccinations Gulf war veterans received, all contributed towards a military "Mass Hysteria." Just as Le Log had been diagnosed by Charcot as suffering from the fear of brain damage, an idea, rather than brain damage itself, Wessely implicated over 100,000 allied soldiers in a force majeure of hysterical symptoms formation.

His position was challenged when, in the June 2007 issue of Radiology, researchers reported that sick Gulf War veterans, when compared with healthy veterans, had 20 percent fewer brain cells in the brain stem, 12 percent less in the right basal ganglia and 5 percent fewer in the left basal ganglia. This brain damage was viewed by researchers to be similar in magnitude to results found in patients with diseases like amyotrophic lateral sclerosis (ALS) and multiple sclerosis, as well as dementia and other degenerative neurological disorders, although the brain areas affected are different.[148]

Crucially the researchers noticed the veterans showed brain-cell damage similar to that seen in the early stages of Parkinson's disease. Just as Sigurjonsson had been faced with seemingly inexplicable "Parkinsonian-type" symptoms in the children he was to later autopsy, these researchers found similar evidence of brain damage in their living subjects. What in the minds of Wessely, Sharpe and colleagues could only be explained through notions of psychosomatic and functional illness had in fact, in some cases at least, a substantial organic basis.

The question must therefore be raised in the current evangelism for hysteria, what pathology are Stone, Sharpe, Bass, Carson, Wessely, Crimlisk and others missing in their diagnosis of "symptoms not caused by disease"? If it is clear that they are not missing the peaks of classic diagnoses, and this is by no means certain, then it is without question that they are missing the diffuse organic basis of the myriad of symptoms they continue to deem to be hysterical.

Charcot's fatal mistake, his bad idea was to assume that an idea of illness could cause the illness itself, that in effect pathology was an optional component in disease, an idea as absurd as claiming that oxygen is an optional part of breathing. "Hysteria can be diagnosed with confidence," claim Stone and colleagues. Yet it is precisely this vivacity that has again and again led medicine into disaster. That what should be based in science and the microscope has constantly strayed into the "art" of sickness rather than the physics of disease. Writing about medicine's failure to learn from its mistakes, E. Hare states:

> *It may be argued that historians ought to pay more attention than they have done to scientific hypotheses which proved to be failures. The trouble with the history of science, and of scientific medicine, is that it has too often been presented as one long success story; whereas,*

> *in fact, a striking feature of the history of science (particularly where science overlaps with medical and social matters) has been the tenacious persistence of supposedly scientific ideas long after they ought to have been abandoned.*[149]

Throughout this book I have constantly made reference to "hysteria" as a diagnosis, a catch all phrase, for the myriad of diffuse and complex pathologies that this group of patients present with[150], a distinct set of symptoms that often remains beyond the ken of contemporary medicine, but to assert that classic diseases such as multiple sclerosis are not being missed, though often correct, does not give the right to assert that no organic pathology is being missed. The startlingly high rate of co-morbidity of medically unexplained symptoms alongside observed organic pathology should be a wake-up call to Alexander Luria's message that "the body is an organism" if one system is slighted then the whole body suffers. Indeed the work of neurologists Ron, Binzer and Kullgren suggests that patients with longer lasting deficits are more likely to have quantifiable pathology alongside their medically unexplained symptoms[151] so that when symptoms persist, patients should be re-investigated. As Jon Donne wrote:

> *No man is an island, entire of itself, every man is a piece of the continent, a part of the main. If a clod be washed away by the sea, Europe is the less, as well as if a promontory were, as well as if a manor of thy friends or of thine own were. Any man's death diminishes me, because I am involved in mankind and therefore never send to know for whom the bell tolls, it tolls for thee.*

Disease presents itself shaped by the community from which we come; we are part of that continent. Yet it also comes from a body whose complex interactions and dependencies are only just being understood. Just because symptoms are medically unexplained, does not mean they are medically inexplicable. Just because neurology sees the signs of hysteria as indicating a lack of organic pathology does not mean that such a view is correct.

If biochemical or structural defects are at the heart of any so called hysterical symptoms then the notion that ideas can trigger illness becomes of minor note against medicine's task of identifying and treating pathology. Scientific explanations often acquire the status of myths as centuries pass, the harmony of the spheres and notions of the body in sympathy with the cosmos, the wandering womb. Such mythologies are important to the arts but they have little relevance to evidence based medicine. The notion of hysteria as a single disease entity in which ideas cause illness is, as Slater recognized, a myth.

When the full force of science is turned to study the human organism time and again as Hyde, Goldstein and Cheney show, as an organism we find "all that becomes a man" in the very heart of matter.[152]

> *There can be little doubt that the term 'hysterical' is often applied as a diagnosis to something that the physician does not understand. It is used as a cloak for ignorance. In addition we can still recognise new neurological diseases. Not only can a patient's symptoms be dismissed as hysterical because the physician makes a mistake out*

of inexperience, but also because the illness has only recently been identified. Neurology has never been and is not static. Many neurological diseases are still not widely recognised ... No doubt there are many other neurological conditions still undiscovered. History tells us that there must be illnesses which presently we do not recognise but dismiss as 'hysterical'.[153] C. D. Marsden

11) Acknowledgments:

This book would never have been written without the support and encouragement of a large number of people. My thanks to Richard Webster in the initial stages of the book and for his mammoth 15 page letter of suggestions for improvement. To Angela Kennedy for her comments on Richard Webster's comments and on the book. My thanks also to Charles Thornley for the many hours of discussion we had on the subject of mental illness, Freud and Jung. I will return all your books and I thank you for the time spent editing this one. To Linda Danielis for her help in editing and generally improving the manuscript and to Juliet for encouraging me to finish what I had started. To Tracy Johnson my phsyiotherapist, to Diane Benson and Yvonne Southam.

To Jon Gay for his efforts in organising my rescue when I fell ill in the Tien Shan. The memory of horses arriving at base camp and Jon nonchalantly trotting his horse towards the tent will stay with me forever. Jon I can never thank you enough. The contribution you made to Kazakh culture was immense; I am told "Mornington Crescent" is now an intrinsic part of Kazakh life. To Fiona for her just about putting up with my focus on this book and criminal neglect of her during its writing. To the staff and patients of the Maudsley who when Fiona was first a patient there for her OCD taught me that psychiatry is indeed a form of control, more like teachers and unruly students than doctor and patient.

To my father also for his many hilarious tales of the psychiatrists he worked alongside. Dad, are you sure these men were not patients in disguise? To Wendy Fox for her recommendation of Prof. Terry Daymond, a superb physician and one who started the process that changed my life. To Dr. Byron Hyde whose respect for his fellow physicians has been my starting point in developing a rational discourse. If I have strayed from this it is to my eternal shame. To the neurologist Dr. Jon Stone who was so deeply kind in answering a number of questions in the process of writing this book. We may disagree on the subject of hysteria but he has informed my understanding when in truth I have been an ignorant amateur. I acknowledge his professionalism and profound dedication to the wellbeing of all his patients. My thanks also to Dr. Vance Spence of ME Research UK and to Prof. Philip James, Dr. Neil Cowley, Dr. Khan, Dr. Sarah Goodman, Dr. Yan Yiannakou, Dr. Nigel Roberts, Dr. Jon Marsh, Dr. Ian Gibson MP, Dr. Julia Newton and Dr. Grainne Gorman of the Royal Victoria Infirmary for taking forward the process of my own diagnosis and treatment away from ignorance and error.

On those two physicians who proved to be in error I will, like Wittgenstein, pass over in silence, this book is for them and their patients.

Index:

Armstrong, Karen, 57
asylum, 82, 113, 124, 127, 131, 133, 144
autonomic nervous system, 134
autopsy, 84, 89, 108, 109, 123, 145
Babinski, 74, 127, 137, 142, 144
basal ganglia, 109, 145
Bass, Chris, 22, 30, 85, 127, 128, 146
Bennett, 51
biomedical, 41
blank spells, 79
brain, 26, 27, 30, 31, 37, 47, 49, 50, 53, 62, 63, 65, 66, 69, 70, 71, 74, 78, 80, 89, 91, 100, 101, 108, 109, 110, 115, 123, 130, 131, 137, 143, 145
Brain perfusion, 109
British Association of Neurologists, 88
British Medical Journal, 121
cardio-vascular, 78
Carson, 96, 99, 146
Carter, Rita, 66, 68
CBT, 56, 57, 126
central nervous system, 2, 21, 31, 47, 48, 64, 69, 118, 134
Charcot, Jean Martin, 13, 38, 39, 40

Charcot, Jean-Martin, 1, 5, 13, 38, 40, 41, 50, 59, 61, 69, 74, 81, 127, 128, 129, 137, 142, 144, 145, 146
Chaudhuri, 112
Cheney, 97, 148
chorea, 27, 40
Chronic Fatigue Syndrome, 97, 121
CJD, 91
Clinical Depression, 109
Combe, Andrew, 26
community, 120, 124, 125, 126, 130, 147
Conversion Disorder, 1, 5
cortex, 80
Creutzfeldt-Jakob disease, 89, 90
Crick, Francis, 31
Crimlisk, Helen, 67, 73, 75, 81, 126, 146
cultural, 11
Dalen, Per, 52
Darwin, 59
Deficit, 2
Dennett, 133
Diagnosis, 5, 26, 28, 30, 34, 39, 43, 53, 55, 61, 62, 65, 69, 70, 71, 75, 76, 77, 82, 83, 86, 88, 90, 93, 94, 97, 104, 110, 114, 122, 125, 126, 127, 129, 133, 134,

136, 137, 138, 139, 143, 144, 146, 147, 148, 162
diagnostic, 13, 22, 35, 43, 51, 58, 61, 78, 86, 88, 110, 138, 142
Diamond, 103
dissociative states, 79
distractability, 63
Doppler, 123
DSM, 55, 134, 136, 137, 144
Ean Proctor, 5, 104, 109, 110, 111
Economo, 63, 64
EEG, 78, 89, 128
electron microscope, 40
encephalitis lethargica, 63, 64
fMRI, 64, 65, 66, 71, 142
Foucault, Michel, 37, 52, 57, 58
Freud, 27, 28, 35, 41, 50, 55, 59, 61, 66, 69, 75, 93, 127, 128, 133, 144, 151
Freudian, 71
functional disorders, 26, 31, 47, 66
Functional Symptoms, 1, 5
Gall and Spurheim, 27
Glover, Mary, 24, 25
Goldstein, 97, 148
Gould, 71
Gowers, 27
haemorrhages, 123
Halligan, Peter, 61, 65, 69
Hansen, Lars, 55, 56
Hartley, David, 23, 26

Hoover
   Dr. Charles Franklin Hoover, 14, 15, 21
Hoover's sign, 14, 15
Hurst, 43, 46, 50, 51, 60, 64, 131, 142
Hyde, 97, 108, 114, 115, 148
hypochondria, 20
Hysteria, 5, 14, 15, 21, 23, 24, 28, 33, 34, 35, 36, 37, 38, 39, 41, 43, 49, 51, 59, 61, 62, 63, 64, 65, 66, 69, 70, 71, 73, 74, 76, 77, 81, 84, 85, 88, 94, 96, 107, 113, 122, 126, 127, 129, 130, 133, 137, 142, 143, 144, 145, 146, 147, 148
iatrogenic, 92, 136
Iceland Outbreak, 115
illness, 9
immune system, 22
industrialization, 124, 132
inflammation, 38
Jackson, Hughlings, 27
Jane, 54, 55
Janet, 47, 127, 137, 142, 144
Jorden, Edward, 24, 25, 26
Kahun Papyrus, 33
Karen Armstrong, 57, 86
Laing, 130
Luria, 74, 147
malingering, 14, 30, 31, 50, 94, 136
Marinacci, Alberto, 56
Marsden, 150

McEvedy and Beard, 107, 108, 115
ME/CFS, 30, 88, 97, 118
Mechanisms, 5, 85
medication, 133
medicine, 9, 13, 21, 22, 23, 28, 31, 33, 34, 37, 48, 50, 51, 52, 54, 58, 60, 69, 73, 77, 80, 86, 88, 97, 120, 122, 127, 129, 132, 138, 140, 144, 146, 147
Mental, 2
Meyer, Adolf, 101
migraine, 27, 41, 52, 64, 69, 78
missed diagnoses, 2, 23, 68, 69, 71, 74, 75, 88, 97, 122
Mitchell, Weir, 74, 118
Morgan-Hughes, 104, 117
motor cortex, 49
Mowbray, 108
MRI, 36, 39, 64, 68, 122, 127
Ms. A, 88, 89, 90, 91, 93
Munchausen's by Proxy, 115
muscle function, 30
Myalgic Encephalomyelitis, 97, 104, 107, 108, 110, 115
neglect, 68, 151
nerve doctors, 38, 133
neurasthenia, 55, 113, 118, 127, 133, 134
neurochemical, 63
neuro-imaging, 66
Neurological, 2, 13, 16, 21, 28, 47, 48, 61, 63, 70, 71, 73, 90, 95, 100, 128, 142, 145, 148
Neurology, 1, 5, 50, 71, 91, 131, 148
neurophysiological, 63
neurotic, 40, 48, 53, 79, 100, 124, 139
neurotransmitters, 60, 66, 134
Newton, Sir Issac, 26
Nichols, 87, 88
Nightingale Rsearch Foundation, 88
organic, 2, 9, 15, 22, 23, 26, 27, 30, 31, 32, 35, 38, 39, 41, 49, 52, 59, 60, 62, 63, 64, 66, 69, 71, 74, 76, 80, 81, 82, 86, 87, 88, 89, 91, 96, 97, 98, 100, 102, 110, 111, 112, 114, 115, 117, 126, 130, 131, 134, 137, 138, 140, 142, 143, 145, 146, 147
Orthostatic, 68
Osler, 8, 97, 131
panic, 77, 78, 79, 113
paralysis agitans, 27
parkinson's, 40
Parkinsonian, 145
pathology, 27, 40, 48, 53, 57, 59, 61, 62, 64, 76, 89, 102, 110, 111, 112, 114, 120, 126, 131, 134, 137, 138, 142, 146, 147, 148
patient, 9, 13, 15, 27, 28, 29, 30, 31, 35, 37, 40, 41, 47,

51, 52, 53, 55, 58, 63, 65, 68, 69, 71, 73, 75, 76, 77, 80, 81, 82, 83, 85, 86, 87, 90, 92, 94, 96, 97, 99, 101, 113, 114, 115, 120, 124, 125, 126, 127, 128, 130, 131, 134, 135, 137, 139, 140, 148, 151
patient advice leaflets, 31
physician, 9, 24, 29, 30, 31, 34, 37, 58, 77, 85, 87, 92, 99, 125, 126, 131, 140, 148, 151
physiological, 23, 97, 115
physiotherapist, 48, 51
physiotherapy, 48, 131, 142
Piet, Dr, 57
placebo effect, 131
plasticity within the cortex, 47
post mortem, 27
Proctor, Ean, 104, 105, 112, 118, 119
Prognosis, 5
pseudo-patients, 83
psychiatric, 9
psychiatry, 31, 41, 50, 51, 52, 57, 58, 60, 77, 82, 88, 90, 92, 96, 115, 117, 120, 121, 130, 131, 144, 151
psychogenic, 70, 71, 72, 96, 127, 132, 134
psychological, 23, 27, 28, 40, 50, 57, 61, 63, 69, 73, 74, 76, 80, 85, 87, 94, 96, 102, 111, 118, 129, 137, 138, 143, 144
psychosis, 27, 75
psychosocial, 11, 85, 92
psychosomatic, 49, 94, 96, 121, 126, 144, 145
QEEG, 39, 69, 78, 115
Ramachadran, 49,81
Raulin, Joseph, 35
religious office, 58
REM sleep, 60
Reynolds, J. Russell, 26
Richardson, 108, 125, 126
Rosenham, David, 82, 83
Royal Free Epidemic, 115
Royal Free Hospital, 115
Ruths, Florian, 55
Sacks, Oliver, 30, 41, 62, 63, 64, 80, 81, 102, 113
Salpêtrière, 13
schizophrenia, 27, 71, 73, 83, 130
Seale Haye, 43
Shakespeare, 58
Sharpe, 5, 41, 90, 92, 94, 96, 99, 100, 101, 102, 108, 116, 122, 137, 138, 140, 145, 146
Shorter, 101
signs, 29
Sigurjonsson, 107, 145
Sir James Paget, 59
Slater, Eliot, 21,76,77,99 148
social, 11
Sophia Mirza, 110

SPECT, 65, 66, 85, 109, 110, 115, 123
Stahl, George Ernest, 36, 37
Stone, 41, 53, 62, 90, 68, 70, 78, 79, 81, 91, 92, 94, 101, 102, 122, 126, 138, 146
Streeton, 68
stress, 21, 22, 50, 61, 79, 103, 112, 126, 144
striatothalamocortical, 66
supernatural, 38
Sydenham, 33, 36, 37
symptoms, 11, 13, 22, 23, 28, 29, 30, 31, 35, 37, 38, 40, 43, 47, 48, 49, 50, 51, 52, 54, 55, 56, 57, 58, 59, 60, 61, 62, 64, 66, 68, 69, 70, 71, 73, 75, 77, 80, 81, 83, 86, 88, 90, 91, 92, 94, 97, 99, 100, 102, 104, 107, 119, 120, 121, 122, 125, 128, 129, 130, 131, 132, 133, 136, 137, 138, 139, 142, 143, 144, 145, 146, 147, 148
Szasz, 95
testing, 2, 11, 83, 126
The Guardian, 54
the Los Angeles epidemic, 115

The McManus report, 119
total body mapping, 88
Treatment, 5, 33, 59
Turkington, Douglas, 58, 60
unexplained symptoms, 13
urbanization, 127
vasomotor, 31, 142
Vuilleumier, Patrik, 64, 65, 66, 68, 69, 85
Wallis, 115, 117, 123
weakness, 14, 21, 22, 28, 33, 40, 52, 61, 62, 63, 68, 69, 70, 71, 74, 77, 80, 82, 86, 103, 133, 134, 139, 140, 142
Webster, Richard, 46, 119, 124, 131, 151, 152
Weil, Simone, 64
Wessely, 5, 96, 99, 105, 108, 109, 110, 111, 112, 114, 115, 116, 117, 118, 144, 145, 146
White, Prof. Peter, 10
Whytt, 33, 34, 38
Willis, 34
Witch Finder General, 58
Wolfe, Dr., 57, 58
Yeats, 79, 96
Ziedler, Martin, 90

12) Glossary

The aim of this short glossary is to provide a brief reference guide for those who are not familiar with some of the technical terms used in this book. It does not aim to be a definitive list of all neurological and psychiatric terms but hopefully should enhance the reader's understanding of some of the issues raised.

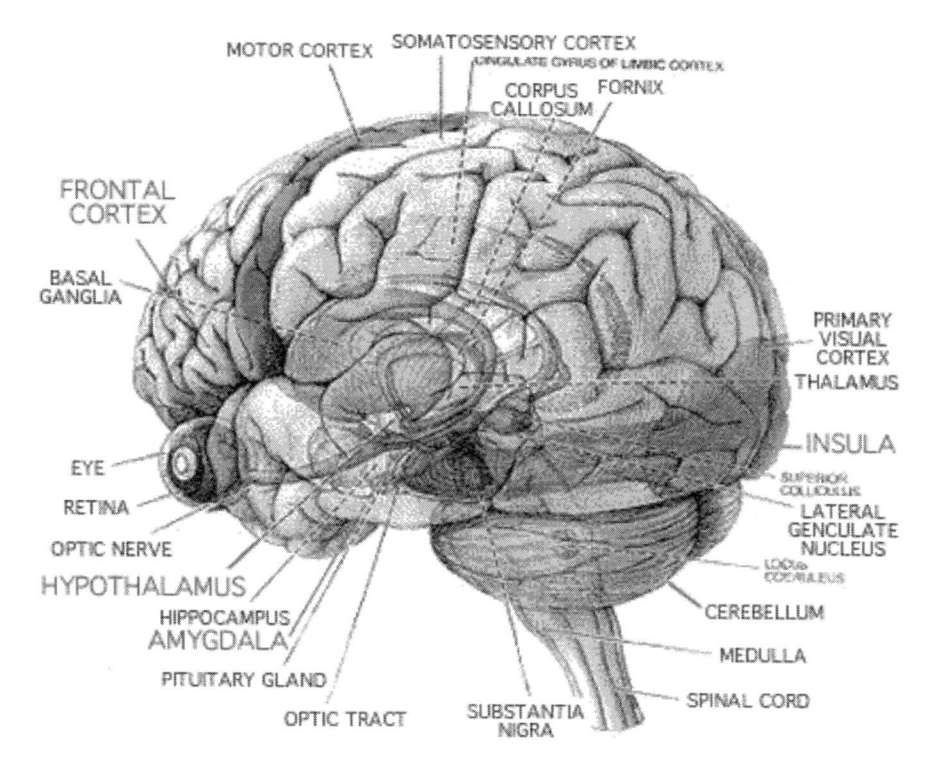

The Major Structures of the Human Brain

Camptocormia: Originally considered a psychogenic disorder, camptocormia, an abnormal posture with marked flexion of thoracolumbar spine that abates in the recumbent position, is becoming an increasingly recognized feature of parkinsonian and dystonic disorders. In other words it was once thought to be "all in the mind" but is now seen as organic

Capillaries: These are the smallest of a body's blood vessels, measuring 5-10 μm, which connect arterioles and venules, and are important for the interchange of oxygen, carbon dioxide, and other substances between blood and tissue cells

Cerebral Perfusion The flow of blood through the brain

CT Computed tomography: originally known as **computed axial tomography** (CAT or **CT scan**) and **body section rentenography**, is a medical imaging method employing tomography where digital geometry processing is used to generate a three-dimensional image of the internal organs

Encephalitis Lethargica: is an atypical form of encephalitis. Also known as sleeping sickness (though different from the sleeping sickness transmitted by the tsetse fly), EL is a devastating illness that swept the world in the 1920s and then vanished as quickly as it had appeared. First described by the neurologist Constantin von Economo in 1916, EL attacks the brain, leaving some victims in a statue-like condition, speechless and motionless.

fMRI: Refers to Functional magnetic Resonance Imagine a technique used to measure blood flow in the brain in response to neural activity

Hemiplegia: A condition where there is paralysis down one side of the person's body

Hemiparesis: A condition in which one half of a person's body is weakened.

Hemodynamic: The flow of blood in the brain as a response to cerebral activity

Hemoglobin: is a protein that is carried by red cells. It picks up oxygen in the lungs and delivers it to the peripheral tissues to maintain the viability of cells.

Hypochondriac: A person who worries excessively about their health

Motor Cortex: The **primary motor cortex** (or **M1**) works in association with pre-motor areas to plan and execute movements.

MRI: Functional Magnetic Resonance Imaging, a scanner used to visualise the brain, similar to x-ray in that it does not measure function

Myopathy: This is a neuromuscular disease in which the muscle fibers do not function for any one of many reasons, resulting in muscular weakness.

Orthostatic: In medicine refers to the state of standing. It is most frequently used to describe the ocndition of orthostatic hypotension that means an unusually low blood pressure when the patient is standing up

Pathognomic: In medicine a pathogniomic sign is a particular sign noted by the doctor that indicates beyond doubt a particular disease is present

Pathology: The study and diagnosis of disease through examination of organs, tissues, cells and bodily fluids.

Phrenology: Is a theory that claims to be able to determine character, personality traits and criminality on the basis of the shape of the head

Physiological: is the science of the mechanical, physical, and biochemical functions of humans in good health, their organs, and the cells of which they are composed. The principal level of focus of physiology is at the level of organs and systems.

Physiologist: A person who studies physiology

Psychoneurosis: This is a catch all term used to describe any mental imbalance that causes distress.

Psychosomatic: **psychosomatic disorder**, now more commonly referred to as **psychophysiologic illness**, is an illness whose symptoms are caused by mental processes of the sufferer rather than immediate physiological causes. If a medical examination can find no physical or organic cause, or if an illness appears to result from emotional conditions such as anger, anxiety, depression or guilt, then it might be classified as psychosomatic.

QEEG quantitative electroencephalogramA brain mapping technique with a higher resolution than standard EEG

Somatic: of the body

SPECT: Single Photon Emission Computed Tomography is a nuclear imaging technique using gamma rays. Can be used to image blood flow in the brain

Visual Transcranial Doppler: The use of ultrasound pulses to image blood flow in the brain and thereby detect any anomalies.

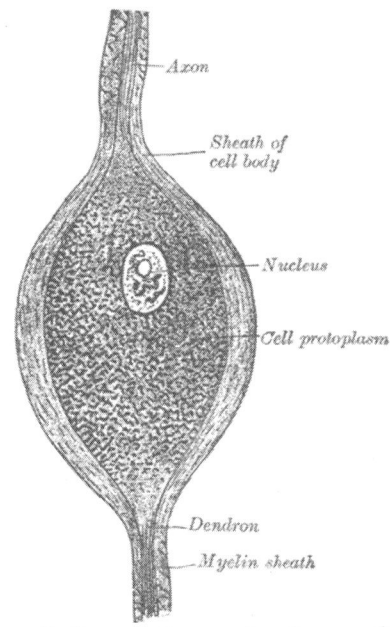

Bipolar nerve cell from the spinal ganglion of the Pike

NOTES:

[1] Illness beliefs and treatment outcome in chronic fatigue syndrome DEALE A. (1) ; CHALDER T. (1) ; WESSELY S. (1) ; Journal of psychosomatic research 1998, vol. 45, n° 1 (91 p.) (27 ref.), pp. 77-83
- "good outcome in ["chronic fatigue syndrome"]is associated with change in avoidance behavior, and related beliefs."

[2] Cognitive behaviour therapy for the chronic fatigue syndrome: a randomised controlled trial
Michael Sharpe, Keith Hawton, Sue Simkin, Christina Surawy, Ann Hackmann, Ivana Klimes, Tim Peto, David Warrell, and Valerie Seagroatt BMJ 1996 312: 22-26

[3] Quoted from Margaret Williams as posted on http://www.meactionuk.org.uk/Whiter_than_white.htm

[4] 2006;6;308-313*Practical Neurology*: Jon Stone Dissociation: what is it and why is it important?

[5] As above

[6] From private correspondence with the office of the Rt. Hon Dr. Ian Gibson MP. Dr. Gibson was quoted by one of his aides hence this is an indirect quote.

[7] FUNCTIONAL SYMPTOMS IN NEUROLOGY: MANAGEMENT J Stone, A Carson, M Sharpe- *Journal of Neurology Neurosurgery and Psychiatry* 2005;76:i13-i21

[8] Prevalence of medically unexplained episodes in frequent attenders categorised by referral complaint: Reid S BMJ 2001;322:767

[9] Proceedings of the British Neuropsychiatry Association Annual Meeting, Institute of Child Health, London, UK, 9–10 February 2006

[10] Stone et al "Hoover's Sign"- Practical Neurology 2001: 1:50-53

[11] personal correspondence

[12] Kahun Papyrus & Also Hippocrates "On Diseases of Women, on Diseases of young girls". Hippocrates also noted that hysteria

tended to occur in older women primarily spinsters who were deprived of sexual intercourse. The precedent for Freud's later theories is obvious.

[13] My own patient notes are a glowing example of the use of "" by the consultant neurologist Dr. Philip Nichols to indicate a psychological rather than an organic problem.

[14] Kahun papyrus and Ebers Papyrus quoted in "Matrons and Marginal Women in Medieval Society" Ed. Robert R. Edwards

[15] Private Correspondence

[16] My thanks to Byron Hyde MD for this simple but superb summary of Freud's contribution to hysteria

[17] Encylopaedia Brittanica online: http://www.britannica.com/eb/topic-586053/telescope

[18] Jay A. Goldstein "Betrayal by the Brain: the neurologic basis of Chronic Fatigue Syndrome, Fibromyalgia Syndrome, and Related Neural Network Disorders" pg 11 Haworth Medical Press

[19] Gould et al 1986 "The validity of hysterical signs and symptoms." The Journal of Nervous Mental Disease 174, 593-7

[20] Eliot Slater, 'Diagnosis of "Hysteria"', *British Medical Journal*, 29 May 1965, p. 1399. See above, Chapter 6, final paragraph.

[21] Conversion Syndrome, Chris Bass, Oxford Medical School Gazette: issue 55 vol2 Part 6

[22] Psychological stress and the human immune system: a meta-analytic study of 30 years of inquiry.Segerstrom SC, Miller GE, Psychol Bull. 2004 Jul;130(4):601-30

[23] Psychol Bull. 1985 Jul;98(1):108-38. Reiew of the effects of stress on cancer in laboratory animals: importance of time of stress application and type of tumor.

[24] Oliver Sacks, "A leg to stand on" Simon & Schuster Adult Publihing Group ISBN: 0684853957. The reader is recommended to the whole book rather than individual references due to the whole been an excellent repost to the notion that hysteria is an

emotional disturbance. Sacks interestingly chooses to make clear he is not a hysteric.

[25] C. David Marsden and Timothy J. Fowler (ed.), Clinical Neurology

[26] Paraphrased from Shaheda et al "Camptocormia" Pathogenesis, classification, and response to therapy" Neurology 2005;65:355-359

[27] Slater revisited: 6 year follow up study of patients with medically unexplained motor symptoms- Crimlisk et al BMJ 1998;316: 582-586 (21 February) also correspondence from Dr. David Bateman, consultant neurologist to author dated September 2007

[28] Harold Merskey- "The Analysis of Hysteria, 2nd edition Understanding Conversion and Dissociation"

[29] *Observations on Man, his frame, his duty, and his expectations.* In Two Parts (1749; 2nd edn, trans. from the German, with *A Sketch of the Life and Character of David Hartley* by his son David Hartley, 1791; 1st edn repr. with an Introduction by Theodore L. Huguelet, Delmar, New York, 1976). Also Allen, Richard C. (1999). *David Hartley on Human Nature.* Albany, N.Y.: SUNY Press. ISBN 0-7914-4233-0

[30] I am indebted to Michael Trimble's article on "Functional diseases" for many of the references in this section: Br Med J (Clin Res Ed). 1982 December 18; 285(6357): 1768–1770

[31] Ferenczi "The Elasticity of Psychoanalytic Technique"- p.91

[32] Freud, quoted in Jones, The Life and Work of Freud, 3: 164

[33] Oliver Sacks, "A leg to stand on"

[34] Bass, Chris, "Does myalgic encephalomyelitis exist" The Lancet Vol. 357 Issue 9721, Pg 1889

[35] Reduced responsiveness is an essential feature of chronic fatigue syndrome: A fMRI study Masaaki Tanaka BMC neurology 2006, 6.9.

editorial note: I am critical of the ue of fMRI in later chapters precisely because it suggests psychological mechanisms as causatory of fMRI abnormalilites. Masakki and his team take the opposite approach.

[36] consultant neurologist RVI, Patient Notes of the author
[37] Patient Advice leaflet on "Functional Weakness" produce by Stone et al for the NHS.
[38] Quoted from "Darwin's Dangerous Idea" by Daniel C. Dennett. The point in quoting is not as Dennett says that models can provide us with insights and therefore Crick is wrong, but rather that Crick is airing the views of many in the field of AI who see such models as *currently* having little resemblance to the organ they attempt to imitate.
[39] Michel Foucault, Madness and Civilisation, chap. 2 "Hysteria and Hypochondria"
[40] Thomas Willis, Essay of the Pathology of the Brain (1684), 69, quoted by Veith, Hysteria, 134
[41] Quoted from Michel Foucault, Madness and Civilisation chap 2. "Hysteria and Hypochondria"
[42] Pg 134 Foucault, Madness and Civilisation, Routledge Classics
[43] The writer Richard Webster asserts that there is no connection between clotting, orthostatic intolerance and seizures, this view is not correct. For a brief overview see:
http://www.emedicine.com/MED/topic3385.htm
[44] Gantz, Katherine *Invention of Hysteria: Charcot and the Photographic Iconography of the Salpetriere (review)* South Central Review - Volume 22, Number 2, Summer 2005, pp. 134-136
[45] Charcot, the Clinician: The Tuesday Lessons : Excerpts from Nine Case Presentations on General Neurology Delivered at the Salpetriere Hospital , translated by Chris Goetz
[46] The following QEEG was analysed by Frank Duffy MD "The father of QEEG", it was performed on a patient who suffered repeated closed head injuries:

EEG DESCRIPTION: In the waking, eyes closed state this 1 hour and 28 minute record reveals 9-11 Hz modestly developed bi-occipital reactive alpha. However, in the fully alert state five bursts, each lasting 1.5-4 sec, of 5-6 Hz medium amplitude theta are observed mostly right occipital (O2) with parietal (P4) spread. These do not appear to be related to drowsiness. They may coincide with abrupt horizontal eye movements which would be consistent with a subtle microsleep event. Eventually, drowsiness is achieved and well developed stage 2 sleep is seen with symmetrical vertex waves and subtle spindles. Prominent rhythmic POSTS are noted, right more than left, normal for age. Intermittent photic stimulation fails to activate discharges. Stimuli used to form evoked potentials do not activate discharges. Hyperventilation shows age appropriate buildup but without activation of seizure discharges. No definite seizure discharges are seen and no clinical or electrographic seizures are observed.

EEG IMPRESSION: A borderline waking and sleep EEG for patient's age due to excessive paroxysmal posterior theta, most marked in the occipital-parietal (O2-P4) region. These appear in the waking states but might possibly represent microsleeps. However, no seizure discharges are seen, no clinical or electrographic seizures are noted, and the waking and sleep background is otherwise within normal limits.

SPECTRAL DESCRIPTION: Spectral analysis of the EEG background in the eyes opened state reveals delta to be maximal in the occipital region, right more than left. Theta is seen to be maximal in the central vertex region. Alpha and beta1 are seen in the occipital region, right more than left. Betas 2 and 3 are artifact dominated. In comparison to an age appropriate normal database from 2-3 Hz there is markedly excessive biposterior 2-3 Hz delta by 5.47 SD which is consistent across three replications. Spectral analysis of the EEG background in the eyes closed state reveals delta to be maximal in the occipital region. There is also excessive bilateral anterior temporal delta suggesting horizontal eye movement. Theta is seen to be maximal in the central vertex region. Alpha in the occipital region, right more than left. In comparison to an age appropriate normal database from 2-3 Hz there is excessive posterior delta, right more than left, maximally 5.20 SD. This abnormality is, once again, consistent across three replications. Alpha is seen to peak, quite broadly from 8.5-11 Hz with mid peak at 9 Hz. Mean for age is 10 Hz. Relative(%) delta is excessive in the occipital region by 2.25 SD. The symmetry function is within normal limits.

SPECTRAL IMPRESSION: An abnormal EEG spectral background due to markedly excessive posterior delta, symmetrically in the eyes opened but slightly more right sided in the eyes closed state. Alpha is seen to maximally peak at a normal 9 Hz, age mean being 10 Hz. Relative(%) spectral data also show excessive occipital delta. Spectral coherence data are within normal limits.

EVOKED POTENTIAL DESCRIPTION: The 4 Hz steady state FMAER shows a bilateral response although it is unusually less well formed over the normally speech dominant left temporal lobe. The AER to click stimuli shows latency delay (N1 by 16 msec and P2 by 40 msec). The P2 is also morphologically distorted. In comparison to an age appropriate normal database from 220-256 msec there is a single epoch of excessive central-parietal positivity by 2.23 SD. This abnormality is consistent across three replications. The auditory odd-ball P300 shows a normal and symmetrical response at CZ and PZ, 19.5 uV at 280 msec. The flash VER is reasonably well formed occipitally with some late time locked alpha. There is an unusual spike-like response especially in the right occipital region at O2. In comparison to an age appropriate normal database from 76-82 msec the SPM shows excessive right occipital focal positivity by 2.27 SD in comparison to the age appropriate database. No additional abnormalities are recognized aside from that easily attributable to the benign, excessive time locked alpha.

EVOKED POTENTIAL IMPRESSION: Probably a normal 4 Hz FMAER but slightly less well developed over the left compared to the right temporal region. An abnormal click AER study due to latency delay, a distorted P2, and one epoch of excessive midline central-parietal positivity. A normal P300 that is normal and symmetrical. A borderline flash VER due to a single epoch of focally excessive early right occipital positivity.

BEAM(QEEG) SUMMARY: The BEAM EEG is borderline in waking and sleep for patient's age due to excessive paroxysmal posterior theta, most marked in the occipital-parietal (O2-P4) region. These bursts appear in the waking states but might possibly represent microsleeps. However, no seizure discharges are seen, no clinical or electrographic seizures are noted, and the waking and sleep background is otherwise within normal limits by visual inspection. Spectral data, however, are quite abnormal due to markedly excessive posterior delta, symmetrically in the eyes opened but slightly more right sided in the eyes closed state. Alpha is seen to maximally peak at a normal 9 Hz, age mean being 10 Hz. Relative(%) spectral data also show excessive occipital delta. Spectral coherence data are within normal limits.The 4 Hz steady state FMAER is probably normal but slightly less well developed over the left compared to the right temporal region. The AER to click stimuli is abnormal due to latency delay, a distorted P2, and one epoch of excessive midline central-parietal positivity. The auditory P300, however, is normal and symmetrical. The VER to flash stimuli is borderline due to a single epoch of focally excessive early right occipital positivity.

Overall, this is a very abnormal study suggestive of an underlying organic
etiology. [1] The O2-P4 paroxysmal theta is of uncertain significance but has often been associated with posterior fossa

difficulties, especially involving the vascular system in older patients. When taken together with the broadly excessive posterior spectral delta, posterior cerebral and/or deep brain abnormality should be suspect. Note, also, the unusually sharp, aberrant VER component at O2 and the 1 Hz slow for age bi-occipital alpha peak. [2] In addition, the FMAER asymmetry raises the possibility for mild left temporal dysfunction. Indeed the abnormal (delayed, distorted) click AER is further suggestive of bitemporal dysfunction. Together, FMAER and AER abnormalities, even when partial, correlate with temporal dysfunction, typically language and memory issues. [3] The curious association of the O2-P4 paroxysmal theta and slow rolling horizontal eye movement that suddenly appears during the apparently awake/alert state is most curious. Such paroxysmal eye movements could signal sudden drowsiness, suggestive of microsleeps, and possibly an underlying sleep disorder. [4] The overall picture does not match our experience with patients having otherwise uncomplicated chronic fatigue syndrome (CFS). Given the clear posterior abnormalities, the bitemporal abnormalities, and possibly deeper abnormalities a broad multifocal underlying process is possible. Head injury is a possible etiology and radiographic evaluation including a view of the vasculature might prove helpful. Various forms of increased intracranial pressure and various demyelinating illnesses and/or small vessel disease may also produce patterns such as noted above. The normal auditory P300 can serve to rule out Schizophrenia (McCarley et al) and severe auditory memory issues. Clinical correlation is advised.

[47] See "the Clinical & Scientific Basis of Myalgic Encephalomyelitis" Ed. Hyde (see 18 below) for a discussion of the use of QEEG in detecting seizure activity in patients with ME. Richardson argued that ME was impossible without seizure activity.
[48] Pg. 23 Parkinson's Disease: Diagnosis and Clinical
Management By Stewart A. Factor, William J. Weiner- Demos medical publishing

[49] Pg. 24 The Analysis of Hysteria: Understanding Conversion and Dissociation By Harold Merskey, Gaskell Academic Series
[50] Pg 24 Parkinson's Disease: Diagnosis and Clinical Management By Stewart A. Factor, William J. Weiner- Demos medical publishing
[51] FUNCTIONAL SYMPTOMS IN NEUROLOGY: MANAGEMENT J Stone, A Carson, M Sharpe- *Journal of Neurology Neurosurgery and Psychiatry* 2005;76:i13-i21
[52] Oliver Sacks, Migraine (1970, revised edition 1992) Paperback, Vintage Books, ISBN 0-375-70406-x
[53] The life and death of Private Harry Farr- Simon Wessely JOURNAL OF THE ROYAL SOCIETY OF MEDICINE, Vol 99 Sept 2006
[54] The life and death of Private Harry Farr- Simon Wessely JOURNAL OF THE ROYAL SOCIETY OF MEDICINE, Vol 99 Sept 2006.

[55] The life and death of Private Harry Farr- Simon Wessely JOURNAL OF THE ROYAL SOCIETY OF MEDICINE, Vol 99 Sept 2006.

[56] See: http://www.wsws.org/articles/1999/nov1999/shot-n16.shtml
[57] Hoge CW, McGurk D, Thomas JL, Cox AL, Engel CC, Castro CA. Mild traumatic brain injury in U.S. soldiers returning from Iraq. N Engl J Med 2008;358:453-463
[58] Mott 1917: 612613, cited in Allan Young, The Harmony of Illusions: Inventing Post-Traumatic Stress Dsorder, Princeton University Press, 1995, p. 51
[59] Webster, Richard: Hysteria Revisited, Unpublished
[60] From personal discussion with various senior physiotherapists based at the YDU West Cumberland Infirmary
[61] Per Dalen- "Somatic Medicine abuses psychiatry- and neglects causal research"- from internet source via google

[62] In conversation
[63] Stone et al, Functional symptoms and signs in neurology: assessment and diagnosis *J Neurol Neurosurg Psychiatry*.2005; 76: i2-i12
[64] Autonomic Dysfunction and Impaired Baroreflex Sensitivity: Implications for Fatigue and Mortality in Primary Biliary Cirrhosis 2006 Julia L Newton, Adrian Davison, Simon Kerr, Nij Bhala, Jessie Pairman1, Jennifer Burt, David E J Jones European Journal of Gastroenterology and Hepatology (in press)

[65] Guardian Unlimited,Tues Feb 20th, Sam Wollaston
[66] Marinacci Alberto quoted from "The Clinical and Scientific basis of Myalgic Encephalomyelitis" (see 18 below) pg 63
[67] CMO Unum Annual Report Oct 2007
[68] Psychminded, September 11 2003
[69] BMC Neuroscience. 2006; 7: 2 Non-rapid eye movement sleep with low muscle tone as a marker of rapid eye movement sleep regulation
Gilberte Tinguely,1 Reto Huber,1 Alexander A Borbély,1 and Peter Acherman
[70] New Approaches to Conversion Hysteria- Halligan et al, BMJ 2000; 320:-1488-1489, 3 June
[71] T. J. Johnson, Senior physiotherapist West Cumberland Hospital, Cumbria
[72] Functional Symptoms and signs in Neurology: assessment and diagnosis Stone et al. J Neurol Neurosurg Pscyhiatry. 2005; 76: i2-i12
[73] "No organic neurological basis" patient notes, Dr. Philip Nichols consultant neurologist RVI, Newcastle-upon-Tyne
[74] Oliver Sacks, A Leg to Stand on
[75] Von Economo C. Encephalitis lethargica. Its sequelae and treatment.
Translated by Newman KO. London: Oxford University Press, 1931.

[76] Slater revisited: 6 year follow up study of patients with medically unexplained motor symptoms- Crimlisk et al BMJ 1998;316: 582-586 (21 February)

[77] See pg491: The clinical and scientific basis of Myalgic Encephalomyelitis: Hyde et al, Nightingale Research Foundation

[78] Theater of Disorder: Patients, Doctors, and the Construction of Illness. By Brant Wenegrat. Oxford University Press, New York 2001. Wenegrat takes the strange view that odd presentations of illness are categorically not instigated by disease in response to cultural pressures. For example if you are having difficulty walking but receive no culturally sanctioned recognition of this you are more likely to conform to the culturally recognized aspects of what a person who has difficulty walking should look like.

[79] Orthostatic Disorders of the Circulation, Mechanisms, Manifestations, and Treatment, Pub.Plenum Medical Book Company, pg180, David H. P. Streeton,

[80] Gould R, Miller BL, Goldberg MA, Benson DF. The validity of hysterical signs and symptoms. J Nerv Ment Dis 1986;174:593-597

[81] Helen L Crimlisk et al Slater revisited: 6 year follow up study of patients with medically unexplained motor symptoms *BMJ* 1998;316:582-586 (21 February 1998)

[82] As above

[83] Luria, A.R. (March 1970). The functional organization of the brain. *Scientific American*, 222(3), 66-78. The Functional Organization of the Brain

[84] Injuries of Nerves and Their Consequences- S. Weir Mitchell (1872)

[85] Eliot Slater, 'What is Hysteria?', in A. Roy (ed.), Hysteria, 1982, p. 40.

[86] Dissociation: what is it and why is it important? Jon Stone Practical Neurology 2006: 6;308-313

[87] Stone stresses that such episodes are commonplace in daily life and can be both good and bad. An acknowledgement that further divorces the actual experience from Stone's insistence that they are also triggered by stress and panic. That dissociation does not occur exclusively in emotional situations further removes it as a useful "sign" of hysteria. See Gould et al also.

[88] This was a common method used by the Maudsley in treating OCD in the early 1990's. 95% of patients who completed their course of treatment had lasting benefit. During the stay by my partner patients were routinely threatened with ejection from the treatment programme. None finished the course which was supposed to continue after discharge, none had lasting benefit

[89] Peter Manu, the psychopathology of functional somatic symptoms pg 1-5

[90] New Scientist 18th Feb 2004 "Wrong Diagnoses are killing patients", Michael Day

[91] Am J Psychiatry 159:528-537, April 2002: "Psychological Versus Biological Clinical Interpretation: A Patient With Prion Disease" H. Brent Solvason, Ph.D., M.D., Brent Harris, M.D., Penelope Zeifert, Ph.D., Benjamin H. Flores, M.D. and Chris Hayward, M.D

[92] Am J Psychiatry 160:391, February 2003, Letter to the Editor Stone et al

[93] The varied neuropscyhiatric presentations of Creutzfeldt Jakob Disease, Psychosoamtics 40:260-263, June 1999

[94] Psychosomatics 45:287-290, August 2004, "Psychosomatic": A Systematic Review of Its Meaning in Newspaper Articles, Jon Stone, Matthew Colyer, Steve Feltbower, Alan Carson, and Michael Sharpe

[95] The Myth of Mental Illness, Dr. Thomas Stephen Szasz

[96] Wessely in interview, ITN "News at Ten", October 1996

[97] Hyde, B., Goldstein, J., Levine, P. (Eds.): *The Clinical and Scientific Basis of Myalgic Encephalomyelitis / Chronic Fatigue Syndrome*. 1992, Nightingale Research Foundation, Press, Ottawa.

[98] Byron Marshall Hyde MD, "The complexities of Diagnosis" Nightingale Research Foundation
[99] Quoted from the Countess of Mar's address in the House of Lords Jan 22, 2004, Hansard
[100] As above
[101] Wessely S: To tell or not to tell? The problem of medically unexplained symptoms, in Ethical Dilemmas in Neurology, vol. 36. Edited by Zeman A, Emmanuel L. London, WB Saunders, 2000
[102] BMJ. 2002 December 21; 325(7378): 1449–1450. What should we say to patients with symptoms unexplained by disease? The "number needed to offend"
Jon Stone, *research fellow in neurology*, Wojtek Wojcik, *medical student*, Daniel Durrance, *medical student,a* Alan Carson, *consultant neuropsychiatrist*, Steff Lewis, *medical statistician*, Lesley MacKenzie, *sister in neurology outpatients*, Charles P Warlow, *professor of medical neurology*, and Michael Sharpe, *reader in psychological medicine*
[103] Virginia T. Sherr MD, Munchausen's syndrome by proxy and Lyme disease: medical misogyny or diagnostic mystery? Medial Hypotheses 2005;65(5) 440-447
[104] Am J Psychiatry 160:1013, May 2003 Letter to the Editor Neurology, Psychiatry, and Neuroscience,JON STONE, M.D. and MICHAEL SHARPE, M.D.
Edinburgh, U.K.
[105] As above
[106] Pg99 A history of psychiatry, Edward Shorter
[107] Oliver Sacks, A leg to Stand on
[108] : Folia Primatol (Basel). 2004 Sep-Oct;75(5):335-8 A case of pseudo-pregnancy in captive brown howler monkeys (Alouatta guariba). Guedes D, Young RJ.

[109] Seth Powser, emedicine, Professor of Psychiatry & Emergency Medicine, Yale University School of Medicine; Medical Director, Crisis Intervention Unit, Section of Emergency Medicine

[110] Based in New York, Dr John Diamond is a founding member of The Royal College of Psychiatrists. In an extract from his recent book (Facets of a Diamond 2003) in the October 2003 issue of the journal "What Doctors Don't Tell You"
[111] Simon Wessely's evidence to Lord Lloyd's enquiry into Gulf War Syndrome
[112] Br Med J. 1970 January 3; 1(5687): 11-15.Concept of Benign Myalgic Encephalomyelitis Colin P. McEvedy and A. W. Beard
[113] Peter D, Moss and Colin P McEvedy, 'An Epidemic of Overbreathing Among
Schoolgirls', British Medical Journal, 26 November 1966, p.1295. The title of McEvedy's article is misleading given that before his involvement the outbreak was thought to be an unusual type of encephalitis and also that one of the teachers also came down with the illness. This ability to overlook facts proved disastrous in their discussion of the Royal Free outbreak where the illness spread to the community.

[114] Byron Hyde, "A New and Simple Definition of Myalgic Encephalomyelitis and a New Simple Definition of Chronic Fatigue Syndrome & A Brief History of Myalgic Encephalomyelitis & An Irreverent History of Chronic Fatigue Syndrome",2007
[115] Peter Manu, the psychopathology of Functional Somatic Syndromes, Haworth Medical Press
[116] pg 106-107 As Above
[117] Dr. Chaudhuri, Neurologist, autopsy report on the death of Sophia Mirza. The fact that Mirza died following lack of hydration and surrounded by "alternative" illness beliefs is often used to downplay Chaudhuri's fundamental finding, findings which in part explain her hypersensitivity to light etc. Symptoms that would otherwise appear psychiatric
[118] Ean's Story- Barbara Proctor
[119] Wessely's statement in his Co-Cure post about the swimming pool allegation and his statement that "For security and safety

reasons there was a CCTV system installed in the pool....nothing remotely like the incident described by Ms Bagnall took place". This was not the conclusion of either report that dealt with Ean Proctor. The first report made no mention of any CCTV videotapes. If video evidence relating to the swimming pool incident had provided clarification, why was it not mentioned in the first report? The second report (the McManus report) stated: "We saw a video film of (Ean) in the water and it appeared to us to be a pleasant and helpful activity". However, the report continues: "We think that it is probable that there was an episode which caused him to be fearful". This indicates that, contrary to Wessely's implication, there was no CCTV of the actual incident in question. Given that Ean was in fact deliberately placed in the water <u>face down</u> with no floating aids whatsoever, it is doubtful if such "treatment" would have been recorded on videotape.- from To set the record straight about Ean Proctor from the Isle of Man- Eileen Marshall and Margaret Williams (pseudonym).

[120] THE MENTAL HEALTH MOVEMENT: PERSECUTION OF PATIENTS? Document prepared for the Countess of Mar by Malcolm Hooper, Emeritus Professor of Medicinal Chemistry, in collaboration with members of the ME community, Department of Life Sciences, University of Sunderland, SR2 7EE, UK December 2003

[121] Byron Hyde, "A New and Simple Definition…" as above

[122] Quoted from Personal Correspondence in relation to a patient misdiagnosed as hysterical but with abnormal QEEG consistent with organic brain trauma

[123] In October 1999 Dr Michael Sharpe (a psychiatrist and prominent member of the Wessely School) gave a lecture at the University of Strathclyde from which this statement is taken.

[124] Statistic reported from AfME survey of 2,338 members (CMO report appendix 2002) 93% CBT had no effect. 79%: reported that they were in severe pain, much or all of the time. 25% ME Group

surveyed 437 members and found 93% found CBT unhelpful. One dutch survey reported that patients who reported improvement were in fact just as functionally disabled.

[125] See: Enteroviral and toxic-mediated myalgic encephalomyelitis- John Richardson, The Haworth Medical Press 2001 for a full discussion of the dangers of exercise on ME patients.

[126] Costa DC Brainstem perfusion is impaired in ME/CFS QJM 1995, 88: 767-73

[127] Hyde also makes reference to this in both the "Complexities of Diagnosis" and the "New Nightingale Definition of ME", Byron Hyde MD

[128] The Proctor's received a letter of support signed on behalf of the Queen by Robert Fellowes. The letter stated "The Queen was so sorry to hear of Ean's illness, but hopes that he will, in due course, make a complete recovery from this debilitating disease. Her Majesty sends her best wishes to Ean and to you and all your family." The use of the word disease is semantically correct as it refers to both disorders of function and structure. In this respect at least Sharpe et al are incorrect in their usage of the word.

[129] THE MENTAL HEALTH MOVEMENT:PERSECUTION OF PATIENTS? A CONSIDERATION OF THE ROLE OF PROFESSOR SIMON WESSELY AND OTHER MEMBERS OF THE "WESSELY SCHOOL" IN THE PERCEPTION OF MYALGIC ENCEPHALOMYELITIS (ME) IN THE UK, Briefing for the House of Commons Select Health Committee- Prof. Malcolm Hooper

[130] Manu, The psychopathology of Functional Somatic Syndromes Pg 1

[131] Systematic review of misdiagnosis of conversion symptoms and "hysteria" BMJ V.331(7523);Oct, 2005 Jon Stone, Roger Smyth et al

[132] A study of the earliest recorded outbreak of epidemic M.E. in the UK appeared in the Thesis of Dr. Andrew Wallis, a Scottish physician, in 1957. It discusses an epidemic in Cumbria in Northern

England in 1955. His definition of the chronic disease included the following features:
A systemic illness with relatively low fever or subnormal temperature.
- Marked muscle fatigability.
- Mental changes with impairment of memory, mood, sleep disorders, depression.
- Involvement of the autonomic nervous system resulting in orthostatic tachycardia, coldness of the extremities, episodes of sweating and profound pallor, sluggish pupils, bowel changes with possible hypothalamic injury.
- Diffuse and variable involvement of the CNS leading to ataxia, weakness and sensory changes in a limb, nerve root or peripheral nerve.
- Muscular pain, tenderness and myalgia.
- Recurrence in some patients over the several years that he followed them.

[133] Enteroviral and toxic-mediated myalgic encephalomyelitis- John Richardson, The Haworth Medical Press 2001

[134] Dr Christopher Bass, Consultant Psychiatrist, Department of Psychological Medicine Oxford Medical School Gazette Issue 55 vol 2, Conversion Syndrome

[135] Effects of skull thickness, anisotropy, and inhomogeneity on forward EEG/ERP computations using a spherical three-dimensional resistor mesh mode, Human Brain mapping- Nichols Chauveau et al, Human Brain mapping, Volume 21 Issue 2 pg 86-87

[136] from private discussion, source: Rev. Ken Kitchin, ex NHS manager. Other discussions with NHS managers reveal that when errors do occur the shredder and a patient's notes often develop an unexpected friendship.

[137] Robert Plomin and Denise Daniels, "Why are children in the Same Family So Different from One Another?" Behavioural and Brain Sciences, 10 (1987), pp 1-59

[138] Hysteria- Richard Webster, unpublished essay, some of which is available on www.richardwebster.net
[139] Death of neurasthenia and its psychological reincarnation. As study of neurasthenia. Ruth E.Taylor, British Journal of Psychiatry (2008) 179, 550-557
[140] Richard M. Fogoros "Dysautonomia: A family of misunderstood disorders"
[141] DSM IV, Somatoform disorders
[142] Somatoform disorders: a help or hindrance to good patient care? Michael Sharpe and Richard Mayou British Journal of Psychiatry (2004) 184, 465-467
[143] Stone et al, What should we say to patients with symptoms unexplained by disease? The "number needed to offend". BMJ. 2002 December 21; 325(7378): 1449–1450
[144] Kessel N. Reassurance. Lancet 1979;i:1128-31.
[145] The Diagnostic Value of Peripheral Vasomotor Reactions in the Psychoneuroses- A.B. Van Der Merwe
[146] A neuroscience of hysteria Matthew R. Broome, Current opinions in Psychiatry 17:465-469
[147] Wessely (in Contemporary Approaches to the Study of Hysteria, Ed. Halligan, Bass and Marshall OUP, 2001)
[148] June 2007 Radiology, Haley and his colleagues
[149] (E. H. Hare, 'Medical Astrology and its Relation to Modern Psychiatry', Proceedings of the Royal Society of Medicine, vol. 70, 1977, pp. 105–10).
[150] For example Kanaan's work using fMRI. What is interesting from Kanaan's work however is not that a patient became weak when asked to recall a traumatic memory but that they remained weak once that recall was over. We are therefore faced with dysfunction in the circuits that manage a physiological response to emotional stress, the "virus" behind the stress induced "cold". Kanaan's work therefore emphasizes yet again the importance of looking for organicity behind unexplained weakness. What must

also be remembered is that such extreme memory related dysfunction is only one part of a broad spectrum of "medically unexplained symptoms".

[151] Patients with long-lasting deficits have more comorbidities (Ron, 1994: Binzer and Kullgren 1998) referenced from "Functional neuroanatomical correlates of hysterical sensorimotor loss" P. Vuilleumier et al Brain (2001) 125, 1077-1090

[152] The words here are those of the Jesuit Teilhard de Chardin whose abiding vision of the continual evolution of consciousness in matter lies at the heart of my own vision of what it is to be human.

[153] C. D. Marsden, 'Hysteria – A Neurologist's View', Psychological Medicine, 1986, vol. 16, pp. 277–88.

Printed in Great Britain
by Amazon.co.uk, Ltd.,
Marston Gate.